A HISTORY OF OREGON ORNITHOLOGY

Great Horned Owl by Susan Lindstedt.

A History of Oregon Ornithology

Territorial Days to the Rise of Birding

Alan L. Contreras
Vjera E. Thompson
Nolan M. Clements

Contributing Writers
Paul Adamus | Jim Anderson | Range Bayer | Ralph Browning |
Danny Bystrak | Jeff Fleischer | Chuck Gates | Carole Hallett |
Hendrik Herlyn | Steven G. Herman | Matthew G. Hunter |
George A. Jobanek | David B. Marshall | Tom McAllister |
Carey E. Myles | M. Cathy Nowak | Mike Patterson |
Noah K. Strycker | David Vick | Teresa Wicks

Art Consultant
Larry B. McQueen

Featured Artists
Junco Bullick | Joe Evanich | Shawneen Finnegan |
Barbara B. Gleason | R. Bruce Horsfall | H. Jon Janosik |
Susan Lindstedt | Larry B. McQueen | Ram Papish |
Elva Hamerstrom Paulson | Emily Poole | Becky Uhler

Oregon State University Press   Corvallis

Library of Congress Cataloging-in-Publication Data

Names: Contreras, Alan, 1956- editor. | Thompson, Vjera E, editor. |
  Clements, Nolan M, editor.
Title: A history of Oregon ornithology : territorial days to the rise of
  birding / editors Alan L. Contreras, Vjera E. Thompson, Nolan M.
  Clements.
Description: Corvallis : Oregon State University Press, 2022. | Includes
  bibliographical references and index.
Identifiers: LCCN 2022027279 | ISBN 9780870714009 (paperback) | ISBN
  9780870714016 (ebook)
Subjects: LCSH: Ornithology—Oregon—History. | Bird—Oregon—History. |
  Birds—Oregon—Identification.
Classification: LCC QL684.O6 H57 2022 | DDC 598.09795/1—dc23/eng/20211216
LC record available at https://lccn.loc.gov/2022027279

∞This paper meets the requirements of ANSI/NISO Z39.48-1992
(Permanence of Paper).

Oregon State University
OSU Press

Oregon State University Press
121 The Valley Library
Corvallis OR 97331-4501
541-737-3166 • fax 541-737-3170
www.osupress.oregonstate.edu

To the memory of Arthur Pope,
first organizer of Oregon ornithology, at age seventeen

Contents

Appendixes

Plates follow page 140.

Acknowledgments

Anyone who spends five minutes looking into the history of Oregon ornithology realizes that the preeminent historian of the subject is George A. Jobanek of Eugene, whose *An Annotated Bibliography of Oregon Bird Literature Published Before 1935* (Oregon State University Press, 1997) is all but a history for that period. His work on the annotated reprint of John Kirk Townsend's journal is also an invaluable resource (see Chapter 2). I am fortunate to have known George—"Chip" to his early friends—since I was a teenage birder in the early 1970s and he wasn't much older. George decided to be a consultant to this book rather than its lead author, but we built much of the early history using his work with permission, as readers will see. His help has been monumental.

Range Bayer's research on early Oregon ornithology was also exceptionally useful, particularly for the Oregon coast and for material on Grace McCormac French. The late David B. Marshall wrote extensively about the history of Oregon ornithology, particularly for the mid-twentieth-century period in which he was most active. We have revised some of his work for use here. Some of it remains in first-person narrative.

Jim Anderson, the late Dave Brown, Ralph Browning, Doug Robinson, Sue Haig, Mike Scott, and Fred Zeillemaker kindly responded to interview questions from Contreras, Clements, David Vick, and Noelle Moen about their own role in Oregon bird study. In-person interviews of Anderson, Scott, and Zeillemaker were made on site in Idaho (by Clements) and Deschutes County, Oregon (by David Vick), and recordings will be archived at Oregon State University.

It is impossible to thank everyone who had a good influence on this project, but we must thank Nancy Brown for recognizing, saving, and passing on the Ruth Hopson Keen field notes, now available in the University of Oregon (UO) special collections. Likewise the late Herb Wisner rescued a significant portion of John E. Patterson's photographs and notes. Herb, age 99 when this book was submitted, gave this material to the editors, and it has now been donated to the Hannon Library at Southern Oregon University (SOU) in Ashland.

Librarians are, as always, an incomparable resource. Emily Johns at the Pacific University library provided Mary Raker photos. Noelle Ebert at Southwestern Oregon Community College helped with information on Ben Fawver. Kate Cleland-Sipfle at the Hannon Library provided historical information from southern Oregon, particularly on Carl Richardson. Local birders Barry McKenzie and Karl Schneck also helped with research on Richardson, Otis Swisher, and Sidney Carter at the SOU library. Rachel Lilley at the OSU Valley Library assisted with the image of Grace McCormac French.

Museum collections were a significant part of our work, both in terms of the historic value of specimens and for the original journals they hold. Pamela Endzweig was especially helpful with Shelton, Bovard, and Prescott material at the UO Museum of Natural and Cultural History, as was Lane Community College student Tye Jeske.

Hope Arnold retyped and edited older material that was deemed suitable for inclusion, particularly that by Dave Marshall and George Jobanek. We are also grateful to the following for their contributions to the book: Dawn Villaescusa helped with historical material for Oregon Birding Association/Oregon Field Ornithologists; Linda Perkins assisted with AOU Checklists; Harry Fuller provided material on David Douglas and Olaus Murie; Mark Nikas loaned Contreras some rare books from the early period of Oregon ornithology; Kay Fagan loaned Lord's 1902 edition of *Birds of Oregon and Washington*; Dona Horine Bolt provided information about Tom McCamant and the Salem Audubon Society; Maeve Sowles summarized information about the Oakridge and Lane County Audubon Societies; Mark Nebeker provided information about ODFW wildlife areas; Sue Haig helped with information about the history of ODFW and made many useful editorial suggestions; Danny Bystrak summarized banding data; Robin Lesher helped with material from *Northwest Science*; Nancy Fraser researched reports on Hubert Prescott's bluebird years; and Robert Warren at the Oregon Historical Society provided assistance with photographs.

Printing of the color plates was funded by a generous grant from the Oregon Birding Association, which also helped underwrite work in preparing this book. Thanks to Kento Ikeda for his excellent index.

The youngest contributor to this project was sixteen; the oldest was ninety-nine, which is a fairly typical picture of the great breadth of interest and activity in Oregon bird study starting in the nineteenth century. A couple of our writers were already dead, a tribute to the staying power of their work. There's history for you.

Every writer needs a Skeptic Tank, the group of people who will read the whole manuscript and point out errors, omissions, and infelicities with a cold eye and a commitment to making a book as good as it can be. Our Tank for this volume included Susan Haig, Hendrik Herlyn, George Jobanek, Mike Scott, and Paul Sullivan. We are particularly appreciative of Haig's many suggestions for improvement of the text, as well as for her willingness to edit *As the Condor Soars: Conserving and Restoring Oregon's Birds* (OSU Press, 2022), which focuses on professional ornithology and conservation work after the middle of the twentieth century. In addition, other readers checked portions of the work. Finally, we benefited greatly from the comments of two external reviewers chosen by the publisher, Daniel Gibson of the University of Alaska Museum and Dave Swanson from the University of South Dakota, both of whom have significant knowledge of Oregon birds.

We are responsible for what we did with their advice, and for the content of this book.

Abbreviations

AOU	American Ornithologists' Union. It merged in 2016 with the Cooper Society to become the American Ornithological Society (AOS).
BBS	Breeding Bird Survey
BOO	Birds of Oregon (listserv)
CBC	Christmas Bird Count
NOA	Northwestern Ornithological Association
NWR	National Wildlife Refuge
OAS	Oregon Audubon Society
OBA	Oregon Birding Association, new name for Oregon Field Ornithologists as of 2012
OBBA	Oregon Breeding Bird Atlas
OBOL	Oregon Birders OnLine (listserv)
OBRC	Oregon Bird Records Committee
ODFW	Oregon Department of Fish and Wildlife
OFO	Oregon Field Ornithologists, established 1980 and name changed to Oregon Birding Association in 2012
OHS	Oregon Historical Society
OSU	Oregon State University
SOU	Southern Oregon State University
SWOC	Southern Willamette Ornithological Club
UO	University of Oregon
USFWS	US Fish and Wildlife Service
USNM	United States National Museum
WFO	Western Field Ornithologists. California-based association including both amateurs and professionals
WFVZ	Western Foundation of Vertebrate Zoology

Introduction

Alan L. Contreras

What is ornithology? In a formal sense ornithology is bird study, from a Greek word meaning "about birds." In practice this term has had a variety of meanings owing in part to the significant role that amateurs have played in adding to knowledge of birds, particularly before the mid-twentieth century and still true in a lesser sense today. In this book ornithology means the *organized study of birds.* For this reason it includes a wide variety of activity from museum and lab study to more diffuse and less formal activities such as Christmas Bird Counts and pelagic trips. This volume focuses on pioneer days, the development of organized bird study in the late nineteenth and early twentieth centuries, and the advent of "citizen science" bird study and recreational birding after about 1960.

Study of birds can and does take many forms. Some who study birds are specialists solely on the tiny parasites that use birds as a host. Some study migration. Yet others focus on particular species or groups of species—for example, Eric Forsman's work on the Northern Spotted Owl or S. Kim Nelson's study of the Marbled Murrelet, both discussed in *As the Condor Soars: Conserving and Restoring Oregon's Birds.* For some observers, nesting behavior is of principal interest, while others study bird sounds or food habits. The list of possibilities is not quite endless, but it is substantial.

As this work progressed to its final stages, it became clear that one volume would be too small to contain all that we wanted to discuss. At that point some existing chapters were peeled off and reassigned to *As the Con-*

dor Soars, which covers academic and professional ornithology and related avian conservation issues from about the 1950s to 2020. The chapters reassigned to *As the Condor Soars* were originally little more than short sketches in some cases, owing to space limits. They were substantially rewritten under that book's senior editor, Susan Haig. As noted above, there is some natural overlap in bird study (e.g., the history of knowledge of Oregon's Yellow Rails, discussed in this volume, is also discussed in *As the Condor Soars*), so readers are likely to find items of interest in both books.

A history of ornithology, unlike a purely scientific text, is necessarily about humans. It is a history of the way humans study birds. It is not a guide to the birds or a set of biographies, yet it has elements of both. Bird observation occupies many people to one extent or another, as we all live in a world filled with birds and notice them from time to time. This book is about people who observed and studied birds in a purposeful, organized way in Oregon, what they did, and, at a basic level, what they learned.

This history is also not a literature review, nor is it intended to be a comprehensive listing of every important piece of work about Oregon birds. Rather, we have tried to present the principal highlights of what has happened with bird study in Oregon in a way that most readers will find of interest. Because it covers modern professional bird study, *As the Condor Soars* has more the tone of a scientific publication, while this volume, though we hope it is thorough and accurate, has a more popular tone. This is intentional.

For those interested in the details of Oregon bird study before 1935, see George A. Jobanek's *Annotated Bibliography of Oregon Bird Literature Published Before 1935* (Oregon State University Press, 1997). If readers are interested in the history of the broader field of ornithology, we suggest Paul Farber's *The Emergence of Ornithology as a Scientific Discipline: 1760–1850* (1982), Erwin Stresemann's *Ornithology from Aristotle to the Present* (1951, English edition 1975), and Mark V. Barrow's *A Passion for Birds: American Ornithology after Audubon* (2000). Other popular histories are available. A unique and rewarding source is Edgar Hume's *Ornithologists of the U.S. Army Medical Corps* (1942, reprinted in 1978). In the late nineteenth and very early twentieth centuries, many of the people studying birds in the American west were in this group.

As you read about the history of bird study in Oregon you will encounter surprising stories about the important role that people without formal training in biology played in expanding knowledge of Oregon's birds. That was true for the first 150 years of Euro-American settlement in Oregon, and, in some parts of bird study, particularly with regard to distribution, it is still true today.

There are also vignettes about people whose appearance in Oregon ornithology is not fully appreciated. For example, Olaus Murie, the famous bird artist and Alaskan explorer, apparently banded the first bird in Oregon (a Brandt's Cormorant). The even more famous bird artist George Miksch Sutton was, as a boy, allowed to arrange the specimens in the University of Oregon museum—until he got kicked out for sliding down the banisters. Jim Anderson, age ninety-one at the time we interviewed him, once spent the night in an eagle's nest. Murie nearly spent the night in Kingdom Come when his boat swamped off Three Arch Rocks in 1914.

This book is loosely structured as a chronological tour of the state's ornithology. Chapter 1 provides a short overview of what bird study was in the period preceding that covered by this volume. It also outlines some themes that appear later in the book. In particular, these relate to two issues of importance in considering what bird study, and ornithology, are. These are the distinction between *amateur* and *professional* study and the difference between *scientific* and *nonscientific* work. These topics are related but different, and ornithology as a field has been one of the principal arenas in which these distinctions have been, and continue to be, discussed.

After that comes the study of living birds and the expanding public interest in bird observation, some of which is now sometimes described (sometimes inaccurately) as "citizen science" and some simply enjoyment of birds, which occupies the rest of the book. This includes the bird reports of early explorers and settlers, efforts to develop and disseminate accurate information about the state's birds, changes in how birds are studied and who studies them, the advent of Audubon societies and other organizations, the twentieth-century separation (and occasional overlap) of professional ornithology and birding and the beginning of electronic data management.

We include essentially no information on aspects of how birds fit into the lives of Native Americans. This is a difficult and complex area of study outside the expertise of the editors and is already covered, at least in part, by other books.[1] It is also a subject that deserves far more space than we could make available. Also, material from early western investigators such as Thomas Nuttall and Vernon Bailey includes some limited information on this topic.

A few chapters are not strictly chronological owing to their subject matter. We chose, for example, to have a separate chapter on museum collections (now discussed mainly in *As the Condor Soars*) because they have been fundamental to much of the ornithological work done in Oregon over an extended period. Likewise, chapters on Charles Emil Bendire, William L. Finley and Herman T. Bohlman, Ira Gabrielson and Stanley Jewett, and David B. Marshall are free-standing mini-biographies, because of the out-sized impact that these individuals had. The chapter on seabirds necessarily extends over many years and shows a blend of the results of amateur and professional bird study.

Much of the book was written expressly for this project, but several sections were adapted from other sources when existing material seemed as good as anything that could be re-done. This includes some of the material on early history (Bendire, Shelton, and the Northwestern Ornithological Association), some material on mid-twentieth-century work and people (Dave Marshall's memoirs in particular) and some later bits as well. We think the variation in tone and focus of these pieces adds some flavor to the story without breaking up the general flow of events. The contributions of George Jobanek, the late Dave Marshall, and Range Bayer to this process were especially useful. It is a tribute to Dave's unique impact on Oregon ornithology that he is credited as a writer in this volume ten years after his death.

Research Tools
For detailed information about the *distribution* and *status* of Oregon birds,

readers are referred to *Birds of Oregon: A General Reference*, by David B. Marshall et al. (Oregon State University Press, 2003), which also covers diet, habitat, and other subjects, and subsequent information in journals such as *Oregon Birds, Western Birds, Northwestern Naturalist, Condor*, and others. Electronic databases such as eBird also provide some current information about Oregon birds, especially seasonality and migration patterns. Marshall et al., in particular, is a superb starting point not just because it contains extensive information but because its four thousand citations provide a road map for further exploration.

The same is true of the extensive bibliography and notes in Gabrielson and Jewett's 1940 *Birds of Oregon* and especially of George A. Jobanek's *Annotated Bibliography*, which serves as a sort of enhanced concordance to Gabrielson and Jewett's book and to this one. We chose to save space in this volume by not replicating many of the photographs in Jobanek's book. Further bibliographic information for the period we cover is available in Scott, Haislip, and Thompson's *1935–1970 Bibliography for Oregon Ornithology* and Mark Egger's update through 1977.[2]

Those interested in finding information about reports of specific birds in Oregon over time can use the eBird or BirdNotes[3] databases for recent records. A useful reference for older records is Clare Watson's *Index to Oregon Bird Reports in Audubon Field Notes and American Birds 1947–1981* (Oregon Field Ornithologists [OFO] Special Pub. No. 3, 1982). Searchable online copies of northwest field notes from these publications (including Washington and many other reports from adjacent states) are maintained on the Oregon Birding Association web site. These can also be searched, albeit more slowly, on the SORA online database. Watson, a professional librarian, also prepared *A List of Oregon Birds, Cross-indexed with Oregon Reports in 'The Season' Section of Bird-Lore* (1986).

The Project
The three editors of this book emerged as the project moved along. Alan Contreras managed the project, recruiting most of the writers and raising funds. Nolan Clements served as a liaison to the professional wildlife man-

agement community, especially the core group at Oregon State University, conducted several interviews, provided library research, and reviewed the entire text, particularly the footnotes. Vjera Thompson was our primary technical editor and "detail person." She reviewed and cleaned up the appendices, rebuilt the timelines, tracked the result of many editorial comments, and communicated with reviewers. The three of us are just about a generation apart, therefore we have distinctly different backgrounds in when we entered the world of bird study. Nolan is a student in zoology, Vjera and I have backgrounds largely outside the sciences. This has been helpful in giving us different angles on what is in the book.

George Jobanek was a sort of "ghost editor" working with us on a wide variety of issues, particularly the years up to 1935 and related photos. His support for this project was essential to its success.

Printing of the color plates in the center was paid for by a grant from the Oregon Birding Association, the state's largest organization devoted exclusively to wild birds. Many thanks to OBA for making these beautiful paintings available. The featured artists were chosen for their significant impact on audiences in Oregon, in particular by illustrating books. In addition, one emerging young artist, Junco Bullick from Portland State University, was included. For reasons of space limits, many other good artists were not included, as this is a history, not a field guide or a show catalog.

State ornithological histories are rare, and we are not aware of one from the United States that has as broad a reach in time and subject as this one, except for Behle's ornithological history of Utah through the 1980s.[4] An excellent history exists for colonial and early statehood Virginia, but devotes only eleven pages to everything in the twentieth century and beyond. A history for Kansas covers only professional academic ornithology.

Contreras finished most of his work on this book thanks in part to a residency at Playa on Summer Lake, for which he is grateful.

We enjoyed producing this book; we hope that you enjoy reading it.

Alan L. Contreras, Eugene
Vjera E. Thompson, Eugene
Nolan M. Clements, La Grande

CHAPTER 1

Origins and Types of Bird Study

Alan L. Contreras

People have observed birds as long as there have been people. Birds have been food, symbols, gifts, representatives of unknown cultures and places, and objects of study. Bird study has origins distant in time and has been as varied as those who paid attention to birds. By the eighteenth century, birds could be said to be subject to a heightened form of observation that slowly resulted in the study of individual birds and of small personal collections.[1] Museums developed in many countries during the eighteenth and nineteenth centuries. Many remain active centers of learning today while some have been absorbed by other organizations or allowed to fade away.

From the early nineteenth century through roughly the middle of the twentieth century, most people considered "professional" ornithologists—a classification rarely seen at the time—were connected in one way or another with the collection and study of bird specimens in what became museum settings. In the United States this meant a relatively small number of institutions in the period at which this history begins. The Philadelphia Academy of Sciences, the American Museum of Natural History in New York, and the US Museum housed in the Smithsonian complex in Washington, DC, were the principal anchors of nineteenth-century ornithology.

Ralph Browning grew up as a "kid birder" in the Rogue Valley of southern Oregon in the 1950s and went on to become one of the nation's leading avian taxonomists, working at the US Museum. He reminds us that ornithology is a unique blend of amateur excitement and professional focus:

> . . . to feed the passion goes beyond birding. Ornithologists study what is of interest to them, usually, unless their funding overseer dictates otherwise. At least the ornithologists crossing my path were mostly following their muse. For example, an ornithologist might find hormone levels in American Robins fascinating whereas another ornithologist may be bored with hormones and excited about geographic variation. I did not choose to write about Yellow Warblers, Yellow Warblers chose me. Like ornithologists, some birders get very excited about shorebirds, or gulls, or sparrows, or all of the above, or desire to know as much as they can about birds in a certain region. —from *Rogue Birder*[2]

This is echoed by Cornell University ornithologist Ken Rosenberg, who noted in a 2019 article that "there's this incredible collaboration between the scientists and the bird-watchers, and it really doesn't exist with other animals and other sciences."[3] This cooperative effort is broadening somewhat owing to online tools such as eBird and iNaturalist. We have attempted to recognize that blend in this book, which is meant to be enjoyable to the reader while providing an accurate, genuine history of bird study in Oregon, our home state.

As we tracked the development of bird study in Oregon, we revisited two important themes: What is science? What is professional? Professional vs. amateur is not a useful distinction when it comes to the generation of valid, useful data. What matters is whether the data are gathered and analyzed in a way that is consistent with accepted scientific methods. Another way of putting this is that what matters is what is done, not by whom it is done. As Robert Pennock has noted, "The fact that empirical evidence, collected and analyzed with proper scientific methods, is the standard, irrespective of the authority of the presenter, is the reason that science is most usefully understood as a methodological practice."[4]

Assuming the presence of appropriate scientific practice, the distinction between authority earned by practice and authority conferred by an external source, nonexistent before the early twentieth century, has become a factor in ornithology since the 1940s. Much early bird study that was considered scientific in its time, and for which claim there is often significant evidence, was conducted by people who either had no formal training or whose training was in another field. Robert Ridgway was essentially self-trained. Elliott Coues was a doctor by training and Charles Emil Bendire was an army medical officer. Stanley Jewett, co-author of *Birds of Oregon* (1940) and senior author of *Birds of Washington State* (1953), had no college degree at all. However, by Jewett's time, bird study was also attracting young men (very few women) who spent time at leading colleges. For example, Alfred C. Shelton and William L. Finley, discussed in later chapters, both attended the University of California at Berkeley.

Ornithology was first taught as a college course at Oberlin in 1895. Marianne Ainley notes that "from the second decade of the 20th century, the number of universities and colleges teaching ornithology increased sharply, producing a corresponding increase in positions for professors, lecturers, and assistants."[5]

TABLE 1. THE PERCENTAGE OF CONTRIBUTIONS BY AMATEUR AUTHORS TO THREE MAJOR ORNITHOLOGICAL JOURNALS

YEAR	THE AUK	THE WILSON BULLETIN	THE CONDOR
1900	72	83	77
1910	75	84	79
1925	63	91	67
1950	47	46	34
1975	12	12	13

Table adapted from Marianne G. Ainley, "The Contribution of the Amateur to North American Ornithology," *Living Bird* 18 (1979–80) 161–177.

The significant increase in professional and academic ornithology in the middle of the twentieth century is obvious in Table 1, which shows the increasing numbers of writers from professional backgrounds in the three best-known bird journals at the time. A survey of these publications today would show an amateur contribution of close to zero, while new journals such as *Journal of Field Ornithology*, *North American Birds*, and *Western Birds* have emerged and continue to publish scientific work by amateurs as well as professionals.

The question of who is or is not a "professional" has been part of bird study since natural history and its predecessor label, natural philosophy, began the dendritic spreading that the sciences display to this day. Paul Farber noted that mid-nineteenth-century bird study included people like Audubon and Bonaparte, who were

> professional in the sense that they were either independently wealthy but expended the amount of time and effort equivalent to a gainfully employed individual, or they were able to support themselves directly or indirectly by their work in ornithology. . . . What held this disparate and widely scattered group together was not a central institution, international society, main publication or founding father, but rather a set of fruitful questions and a common general goal.[6]

Farber, who considers the term "professional" to be problematic if applied prior to 1850, also points out that the mid-nineteenth century was a time when ornithology shifted "from the salon to the study," and became both more international and more disciplinary, as distinguished from simple personal interest.[7] Erwin Stresemann noted that

> at the beginning of the 1920s, ornithology changed fundamentally. Up to that time, anyone could be regarded as an expert who was well acquainted with systematics, distribution, and "habits." . . . Most representatives of "scientific zoology" viewed ornithology as the province of amateurs, whose findings could not mean much to researchers into causation. The developments between 1920 and 1950 changed all that.[8]

By this, Stresemann meant that bird *study* was expanding into bird *science*, with many of its problems and techniques beginning to overlap with those used in other sciences. The flip side of this change was that research into birds became a legitimate part of sciences that had heretofore treated ornithology as a sort of strange cousin.

One new profession, that of fish and wildlife management, grew from this newly fertile ground. In Oregon, David B. Marshall, whose career is covered in a later chapter, was part of this new program at Oregon State University shortly after his service in World War II. This field can clearly be considered a profession and is an applied science that absorbed a certain segment of people interested in studying and working with birds as their principal job. The first woman to earn a PhD in this field was Frances Hamerstrom (Iowa State), who became an expert on grouse and raptors. We take special pleasure in mentioning her because her daughter, Elva Hamerstrom Paulson, is a well-known Oregon bird artist whose work appears in this book and in *Birds of Oregon* (2003).

This transition period will make many appearances in this book because it overlapped with early bird study in Oregon: roughly the hundred years leading up to the mid-twentieth century. The kind of work done by Frank Chapman, who left school at age sixteen to become a bank clerk yet spent fifty-four years at the American Museum, would, by the mid-twentieth century, be considered professional but not within the mainstream of science, in part because science itself was changing.

Others such as pioneering ethologist Margaret Morse Nice would be in the opposite situation: her work was widely respected for its research standards (Konrad Lorenz called it "the first long-term field investigation of the individual life of any free-living wild animal"), but she conducted her enormous Song Sparrow study in the 1930s in her spare time and was thus not a "professional." Nonetheless, Milton Trautman's memorial for Nice in the *Auk* mentions that the leading American ornithologist Ernst Mayr wrote to him stating that Nice "almost single-handedly initiated a new era in American ornithology and the only effective countermovement against the list-chasing movement." Other examples of scientific work by non-professionals include Lawrence Walkinshaw's studies of Sandhill Cranes, Harold Mayfield's work on Kirtland's Warbler, and Louise de Kiriline Lawrence's woodpecker studies published by the American Ornithologists' Union (AOU).

Steve Howell, author of major reference books on gulls, molt, and hummingbirds, and his co-authors note that "in the 1950s there was relatively little recreational birding and most new records were found and collected by field ornithologists."[9] A review of publications such as *Condor* during the 1940s through the 1960s shows this to be correct. In subsequent years, as academic ornithology moved away from questions of distribution, the new field observers that became known as the birding community increasingly assumed this role. That said, many who held full-time positions studying birds also enjoyed observing birds in a more casual way, while some people who were not so employed have made significant contributions to knowledge of birds. Thus the question of "who is a professional" may ultimately be unanswerable. Perhaps that suggests that the word is less important than it seems. This volume takes us through and beyond the point at which these changes in roles and expectations happened, following the fork of the birding community. The companion volume, titled *As the Condor Soars*, follows the development of the academic and professional bird study community, particularly in its work on important conservation issues.

For further information on this period and the overlapping roles of amateurs and professionals, see Mayr's epilogue in the English edition of Stresemann's history of ornithology cited herein.

CHAPTER 2

Euro-American Land-based Expeditions, 1804–1859

Alan L. Contreras

Rumors and Fizzled Attempts • *The Lewis and Clark Expedition* • *John Kirk Townsend* • *David Douglas* • *Pacific Railroad Surveys* • *John S. Newberry* • *James G. Cooper* • *George Suckley*

Prior to the iconic expedition of Lewis and Clark in 1804–1806 there had been a number of vague and generally inconsequential rumors, thirdhand reports and false starts related to the desire of Euro-Americans in eastern North America to learn about the west. Many of these preliminary sputterings revolved around two main issues, the desire for an expanded fur trade and an interest in determining a hypothetical Northwest Passage.

In a later chapter we discuss birds noted in explorations by sea, but for now suffice it to say that seafaring exploration prior to colonial days tended to discover useful information to the south and the north of what we now call Oregon and Washington, leaving the coastlines of those two states largely shrouded in mist.[1] The great bulk of early bird knowledge in Oregon came from the overland expeditions.

Lewis and Clark

It is hard to overstate the sheer scale of the Lewis and Clark expedition: it has been called the "Corps of Discovery," a title never since applied to an

American expedition and now perhaps held in long reserve, a sort of suspended animation, for the eventual exploration of space. The parallel would be fitting. Standard texts on the expedition are many and varied and seem to cover everything from boot-nails to native relations to trees.

For our purposes, ornithology in Oregon by non-natives begins on Saturday, November 2, 1805, when the following entry was made in the expedition journal for a site twenty-seven miles below the Cascades of the Columbia:

> We saw great numbers of water-fowl, such as swan, geese, ducks of various kinds, gulls, plover and the white and gray brant, of which last we killed 18.

Ira Gabrielson and Stanley Jewett in *Birds of Oregon* (1940) speculate that

> this "swan" might have been either the Trumpeter or Whistling Swan, both of which wintered on the Columbia in the early days, the "plover" was undoubtedly the Killdeer; and the "geese" were without doubt the Lesser Snow Goose and the Lesser Canada Goose, the latter form being the one that still frequents the river in numbers at that season.[2]

So began what we consider the Oregon State List of birds. Lewis and Clark made several other additions to this list later in their expedition. On November 30, 1805, near Astoria they reported, among other animals,

> a few black ducks, of a species common in the United States, living in large flocks, and feeding on grass, they are distinguished by a sharp white beak, toes separated, and by having no craw. Besides these wild fowl, there are in this neighborhood a large kind of buzzard with white wings, the gray and the bald eagle, the large red-tailed hawk, the blue magpie, and great numbers of ravens and crows. We observe, however, few small birds, the one which has most attracted our attention being a small brown bird, which seems to frequent logs and the roots of trees.

Thus did American Coot, California Condor, Red-tailed Hawk, Steller's Jay, Common Raven, American Crow, and Pacific Wren join the list.

On January 2, 1806, the expedition added Sandhill Crane, cormorants, Canvasback, Mallard, Canada Jay, definitely reported both swans and again noted "the beautiful buzzard of the Columbia," another report of condor. Downy Woodpecker was added in April and brown sparrows were reported throughout this period. The last report containing an identifiable new species was on April 9, 1806, when two Turkey Vultures were seen.

Pacific Wren by Joe Evanich.

What was the true range of the California Condor in the early nineteenth century? There is no universal agreement on the answer, but some things are fairly clear. The condor ranged north to the Columbia Valley, mainly along the coast and Columbia River, where some reports have it feeding on salmon along Columbia cascades that no longer exist, such as Celilo Falls. It also occurred at least occasionally north to southern British Columbia. It is likely that the condor has been declining for thousands of years, as there are bones from as far away as the Gulf Coast and New York. Oregon has been a leader in condor research and repropagation. See *As the Condor Soars* for a discussion of these efforts.[3]

David Douglas

The pace of westward exploration was remarkably slow for the next few decades, and thus not much was added to knowledge of Oregon birds by non-natives. Scottish botanist David Douglas, whose name graces the Douglas-fir, landed near present-day Astoria on April 9, 1825, after almost two months lying-to offshore waiting to pass the notoriously nasty Columbia River bar during a period of late-winter storms. He remained in the Northwest for two years, adding an excellent description of Band-tailed

Pigeon, some better data on the range of California Condor ("nowhere so abundantly as in the Columbian Valley between the Grand Rapids and the sea"). He is remembered today mainly for his comments on the condor, the abundance of various Galliformes, including Sharp-tailed Grouse, and for his ever-ambiguous statement as to the range of the White-tailed Ptarmigan in the Northwest:

> On the north-west coast it exists as low as 45° 7', the position of Mt. Hood. This is the same bird as the Scotch Ptarmigan.

As the species has never since been reported on Mt. Hood, but only from the Goat Rocks region closer to Mt. Rainier, and northward, Douglas's statement is generally considered to be a conflation of reports and locations that do not belong together.

Meriwether Lewis's sketch of the head of a condor. There are few if any textual images of an Oregon bird more iconic than that of the "Beautiful Buzzard of the Columbia." The last generally accepted report was from 1904 in the Umpqua Valley, though there were plausible reports into the 1920s. The definitive source for California Condor information for the Northwest is Jesse D'Elia and Susan Haig, California Condors in the Pacific Northwest *(Corvallis: Oregon State University Press, 2013). For an interesting and speculative discussion of condors in the region, particularly the oral commentaries of Northwest tribes, see also Brian Sharp, "The California Condor in Northwestern North America,"* Western Birds *43 (2012): 54–89. For another view of condors in the context of the Columbia region, see Jack Nisbet,* Visible Bones: Journeys Across Time in the Columbia River Country *(Seattle: Sasquatch Books, 2003).*

White-tailed Ptarmigan from Washington and Colorado were intro-duced in the Wallowa Mountains in 1967–1969. Isolated reports contin-ued in that region until the 1990s, and their status in northeast Oregon is currently unknown, although they are assumed to be extirpated. Rumors of birds on Mt. Hood occurred shortly after the eruption of Mt. St. Helens and there have been other isolated reports.[4]

John Kirk Townsend

The elapsed time between John Kirk Townsend's arrival in Oregon in Au-gust 1834 as a member of Nathaniel Wyeth's expedition and Oregon state-hood in February 1859 was only twenty-five years, an astonishing realiza-tion given the difficulty of travel to and in the Northwest. Into those years was compressed most of the Oregon Trail migrations, including the Great Migration of 1843. Our narrative is concerned with birds and those inter-ested in them, and Townsend was the first white visitor to come to the re-gion largely as a naturalist, collecting birds and other wildlife intentionally as part of a scientific enterprise. In that sense he was both a professional and a scientist, though his record-keeping was rather patchy.

The Wyeth expedition, which was followed for part of its route by well-known Oregon missionary Jason Lee, who was engaged in a cattle drive, probably entered Oregon on August 23 or 24, 1834, somewhere near present-day Adrian, Malheur County, as that was the standard crossing point on that trail. They stopped midday on the 24th on the banks of what they called Malheur's Creek, named by Peter Skene Ogden's trapping par-ty in 1825–1826 after its cache of furs was stolen. Townsend's group fol-lowed the Malheur northeastward back to the Snake River before the party split, with some heading out to trap and others going directly over the Blue Mountains and down the Columbia.

Those he traveled with, even botanist Thomas Nuttall, were not always fully attuned to the priorities of ornithology. Townsend writes:

> Having nothing prepared for dinner to-day, I strolled along the stream
> above the camp, and made a meal of rose buds, of which I collected an
> abundance; and on returning, I was surprised to find Mr. N and Captain

T. picking the last bones of a bird which they had cooked. Upon inquiry, I ascertained that the subject was an unfortunate owl which I had killed in the morning, and had intended to preserve, as a specimen. The temptation was too great to be resisted by the hungry captain and naturalist, and the bird of wisdom lost the immortality which he might otherwise have acquired.[5]

The record does not tell us whether Townsend ever ate any of Nuttall's plant specimens. Townsend did, however, do a significant amount of ornithological collecting, portions of which were later used by Audubon, who painted images from seventy-four of Townsend's specimens, though Townsend thought Audubon, whose actual knowledge of birds was quite limited, misused some of the notes accompanying the specimens. The bulk of this collection happened in the lower Columbia region in the vicinity of Fort William on present-day Sauvie Island. The fort was initially sited near "the mouth of Multnomah" but later moved to a point across the channel from the beginning of the Logie Trail access point.[6] This was not a US military fort and was quickly abandoned.

Townsend also collected in the Willamette Valley, west to Astoria (the unexpected type locality for the eponymous solitaire) and as far east as the Blue Mountains. He did not limit his time to collecting birds—he also plundered native burial canoes for skulls sent to a phrenologist friend, was appointed surgeon to Fort Vancouver by John McLoughlin (succeeded by his friend William Tolmie, original namesake of what we now know as MacGillivray's Warbler), and served as magistrate in a murder trial.

For those interested in reading more about Townsend's experiences during and after the Wyeth expedition, see George A. Jobanek's annotated edition of Townsend's *Narrative of a Journey Across the Rocky Mountains to the Columbia River* (Oregon State University Press, 1999).

Townsend's name is memorialized today by a warbler, a solitaire, and several mammals. He reported 209 then-recognized bird species in "Oregon." Of these, he was the first to report 28, of which 24 have descriptive material available—he was not always the most accurate record-keeper. This can be viewed as the beginning of systematic study of Oregon birds for scientific purposes, though some were first found outside the state's present-day boundaries.

The bird list includes those that we would consider subspecies today, or that can't be identified at all, but it also includes the first records known to science of Chestnut-backed Chickadee, Bushtit, Sage Thrasher, MacGillivray's Warbler, Hermit Warbler, Black-throated Gray Warbler, Townsend's Warbler (named by Nuttall), "Audubon's" Warbler, Western Bluebird, Townsend's Solitaire (named by Audubon), Vaux's Swift, Black Oystercatcher, Mountain Plover (on the plains *en route*), and Pelagic Cormorant. Most of these were from present-day Oregon or from the Washington side of the Columbia River. For detailed notes on type specimens found in Oregon, see M. Ralph Browning, "Type Specimens of Birds Collected in Oregon."[7]

One notable aspect of Townsend's list of "firsts" is how many of them are of small birds. Until Townsend's collecting, western explorers in North America were focused on survival and daily progress. Large birds, ideally edible, were important. Bushtits were part of the dust of the trail. Townsend, once he arrived on the Columbia, had what amounted to a base of operations from which he could collect as he wished. He was also therefore the first "Oregon" collector in the sense of living here for an extended period and focusing on the birds of the region.

One of Townsend's guides on his travels was Baptiste Dorion, son of Marie and Pierre Dorion, who had been part of the near-disaster Hunt overland party sent out by John Jacob Astor in 1810–1812.[8] A depiction of Marie, a Native woman, appears today in a mural in the Oregon State Capitol in Salem.

Birds of the Pacific Railroad Surveys

As the United States expanded westward, railroads became the transportation of choice by the early nineteenth century. Already in widespread use in the east, the push in pre–Civil War years was westward. The US government sent out expeditions in 1853 and 1855 to determine the best possibilities for a transcontinental route, the first under Lt. Robert Williamson. The second expedition was in 1855, when Williamson was joined by Lt. Henry L. Abbot of the Corps of Topographical Engineers and naturalist John S.

Newberry, who also served as assistant surgeon. Their report was published in 1857 in volume 6 of the Pacific Railroad Reports under Abbot's name.

Owing to staff changes as the expedition moved north from California to and through Oregon, Newberry himself was not always present, and some of the naturalist duties fell to C. D. Anderson. Newberry did write the final report that appeared in 1857. It includes 173 species of birds, including Lieutenant Williamson's eponymous sapsucker and the "common" Sharp-tailed Grouse—a bird now extirpated from Oregon except for a few reintroductions.

It also includes references to Nuttall's Woodpecker and Black Phoebe in southwestern Oregon that are, at first, quite plausible, unless you know the history of the expedition. In his annotated bibliography of Oregon bird publications before 1935, George Jobanek explains:

> The Nuttall Woodpecker ... [Gabrielson and Jewett wrote] "Newberry collected in the Umpqua Valley ... in August 1855." However, in August 1855, the survey party was in the Klamath Basin. Furthermore, Newberry himself never visited the Umpqua Valley, having left the party prior to the march through this valley south. In his report, Newberry merely noted the Nuttall's Woodpecker as "common in California."
>
> [Lester] Short found the specimen to which Gabrielson and Jewett undoubtedly refer ... labeled "Umpqua Valley, Newberry" but dated in the museum catalog as November, 1855. ... Short points out that the specimen, a female, "is in fresh fall plumage showing slight effects of wear and no indication of molt ... so it was undoubtedly secured in the fall rather than in August." ...
>
> The survey party crossed the border into California on Nov. 6, 1855. It seems unlikely that Newberry himself collected the bird, since he presumably would not have mislabeled it as collected in the Umpqua Valley, not having visited the region.[9]

Essentially the same issue arises regarding the Black Phoebe specimen. Thus it is likely that this early report of these species in Oregon is actually referable to northern California.

The other railroad route explored in Oregon was more northerly, its bird records having been collected primarily from today's Wasco County. This was George Suckley's work. Some of what Suckley collected happened after he left the survey. His report contains considerable speculation about birds that "might" occur in eastern Oregon, but also notes some specific records, for example what "seemed to be" an immature Red-throated Loon collected near Fort Dalles on March 20, 1855.

Spencer Baird, John Cassin, and George Lawrence published a summary of all of the bird data from the Pacific Railroad surveys in 1858.[10]

CHAPTER 3

The Contributions of Charles Emil Bendire

M. Ralph Browning, Hendrik Herlyn, and George A. Jobanek

One of the first of the US military officers (often doctors) to study birds is also one of the most well known today, Charles Emil Bendire. He has a geographic feature named after him, Bendire Mountain, north of Juntura, Malheur County, that is further memorialized by Ada Hastings Hedges's poem "Desert Mountain."[1] It has a darker additional memorial in the Bendire Fire of August 2015 in the same region.

While posted at Camp Harney and Fort Klamath in the 1870s and 1880s, Bendire made the first investigations of southeastern Oregon bird life. Ira Gabrielson and Stanley Jewett, in *Birds of Oregon*, noted that Bendire's Oregon bird studies "probably make the greatest individual contribution to the knowledge of the birds of the state," perhaps exceeded since only by Jewett's own efforts.[2]

Charles Bendire began life as Karl Emil Bender on April 27, 1836, in König im Odenwald in the Grand

Charles E. Bendire. Photo courtesy of the US National Museum.

Duchy of Hesse-Darmstadt (a town now known as Bad König, thirty miles southeast of Frankfurt, Germany). His father was a forester, and probably through accompanying him, young Karl became intimately familiar with the German landscape. "It was amid the scenes and surroundings of his boyhood days," a writer for *Forest and Stream* magazine asserted, "that those tastes were imbibed for the study of natural history which were destined later to dominate his life."

In 1853, upon the advice of a friend, Bender sailed for New York. On June 10, 1854, at the age of eighteen, he changed his name to Charles Bendire and enlisted as a private in the United States Army. He served at military posts in New Mexico and Arizona and later in the Army of the Potomac during the Civil War. After the war, he was promoted and served in Louisiana and California, and then was posted to the Pacific Northwest at Fort Lapwai, Idaho. Fort Lapwai was where Bendire began the scientific work and collecting that would make him renowned as a naturalist and oologist. He later served in Arizona and Idaho; in Arizona he corresponded with Joel A. Allen, Spencer F. Baird, and Thomas Brewer. In particular, he began an extended correspondence with Elliott Coues. Coues named Bendire's Thrasher after him, the type having been collected by Bendire. In November 1874, he arrived at Camp Harney, in southeastern Oregon, having been promoted to captain.

Camp Harney had been established seven years before as a base of operations against Paiute Indians in southeastern Oregon. Located about twelve miles east of the present-day city of Burns, the camp was situated, as Bendire described it:

> on the southern slope of one of the western spurs of the Blue Mountains of Oregon, at an altitude of about 4800 feet.... To the north of the post the country is mountainous and generally well-timbered with forests of pine, spruce and fir, and groves of aspens and junipers; in all other directions it is almost destitute of trees of any size.
>
> South of the post, the sagebrush flats stretched for miles, the landscape being "fully as desolate, if not more so, than the worst part of Arizona." But there was water here, Lakes Malheur and Harney and

many smaller lakes and ponds, hot and cold springs, and while most held brackish water, the chain of water formed "a natural highway and convenient resting places for the immense hordes of water fowl passing through" during the migrations.

The camp was centered in a vast corner of the state completely unexplored ornithologically. John K. Townsend passed through to the north and east. John S. Newberry and Elisha Sterling traversed the eastern slope of the Cascade Mountains; they traveled through the Klamath basin, but their search for an economical railroad route to the Columbia River never led them into this southeastern corner of the state. John C. Fremont, on his expedition of 1843–1844, skirted the corner on his march south through the Summer Lake basin, but the ornithological observations resulting from Fremont's expedition were negligible.

Bendire, having developed his ornithological skills at nearby Fort Lapwai and refined them at Camp Lowell, Arizona, now had an opportunity to be the first ornithologist to explore the region. He began his ornithological investigations in the vicinity of the camp, sometimes accompanied by another officer interested in birds, Lieutenant George R. Bacon. One bird that soon caught his attention was a large whitish woodpecker, or so it appeared to be. He soon discovered, however, that it was not a woodpecker at all but what we know as Clark's Nutcracker, and he diligently endeavored to discover its nest and collect the eggs.

He recounted his efforts in letters to Joel A. Allen and Thomas Brewer. Allen, editor of the *Bulletin of the Nuttall Ornithological Club*, persuaded Bendire to let him compile his letters about the nutcracker into an article about the species' breeding habits. This became the first article published under Bendire's name, appearing in 1876. The year before, Thomas Brewer published under his own name some of Bendire's Harney notes.[3] As a first report of the birds from southeastern Oregon, despite the fact that it covered a period of only seven months, November 1874 to May 1875, this was a significant contribution to western ornithology.

Bendire spent most of his time in the vicinity of Camp Harney, but when he was able, he traveled south to the large lakes of Harney and Mal-heur on duck hunting and egg collecting expeditions. In the April after his arrival, Bendire made two visits to nesting colonies of American White Pel-ican, Double-crested Cormorant, and Great Blue Heron at Malheur Lake. The visits were taxing. "A protracted stay on the island," he wrote in the journal *Ornithologist and Oölogist* (published in 1882), "was anything but pleasant, the whole place being alive with fleas, and the stench from de-caying fish was almost unbearable. The young [pelicans], none of which seemed to be a week or so old, were perfectly naked, not a sign of a feather being visible, and they could not be called attractive creatures."

Bendire took eggs for his collection, and also brought fertile pelican eggs back to the camp and placed them under a domestic chicken; they hatched in 29 days. "The injured and disgusted look of that poor bird at the result of her lengthy and protracted setting haunts me still."

Bendire's Red-shouldered Hawks

One of the most interesting discoveries made by Bendire was doubted for scores of years simply because it seemed unlikely. Gabrielson and Jewett placed the Red-shouldered Hawk on the Oregon hypothetical list because they doubted the validity of the identifications of O. B. Johnson (1880) and Bendire. Johnson had recorded the species in the Willamette Valley and mentioned a specimen of it. Nothing further is known about this specimen.

The record of Bendire is based on three eggs he collected in April and May 1878 at two nests near Camp Harney northeast of present-day Burns in what is now Harney County. He ascribed these to *Buteo lineatus elegans*. These three eggs are now in the collection of the US National Museum. Ralph Browning compared the eggs to those of Red-tailed, Swainson's, and Broad-winged Hawks as well as to Red-shouldered Hawks and found them to belong to Red-shouldered Hawks as Bendire originally stated. These eggs have also been examined by A. C. Bent and Herbert Friedmann, both of whom agree with Bendire's original identification.

Further evidence for the occurrence of Red-shouldered Hawks in Ore-gon in the 1870s comes from Bendire's field catalogue. On May 6, 1878, he wrote, "Nest in a tall juniper thought at first they were a pair of Swainson's Hawks, but they are bright chestnut red all over lower parts. The nest was a mile from the other found, a mile higher up Archies Creek and as it was so far off I took the single egg along. This bird is scarce about here."

Bendire collected a female, according to his field catalogue, on April 17 when the first nest was visited, but the present location of this specimen is unknown. Bendire's description of the birds on May 6, 1878, and the three extant eggs provide ample evidence that *Buteo lineatus* actually occurred in Oregon, and as a breeder, prior to the next known record on April 10, 1976, at Harbor, Curry County (OBRC). Since 1976 the species has spread widely in Oregon and probably breeds occasionally in Harney County, where Bendire first found it almost a hundred years earlier.

Finally, it may be worth noting that John S. Newberry in 1857 reported the Red-shouldered Hawk as "common in those parts of northern California and eastern Oregon traversed by our party." This suggests that the species contracted its range not long after Bendire's time, only to expand again in the late twentieth century.[4]

In 1877, Bendire published additional material on the birds of the Camp Harney region.[5] Much of Bendire's contributions to Oregon ornithology reside in this paper and the earlier paper facilitated by Brewer. Bendire listed 186 species on his main list, observations being principally from Camp Harney and Malheur Lake.

One species, the White-faced Ibis, was included based solely on Nevada records and that it was "presumed that it ranges into" Oregon). It remained for Henry Henshaw to find the first birds, in the Warner Basin in 1879.

Bendire prefaced his species accounts with the following explanation, including his vivid description of an area now well known and dear to many Oregon observers:

This list is not given as a complete exponent of the avifauna of Southeastern Oregon. I am well aware that there still remain many species to be added, particularly of water birds. As far as it goes, it has been compiled from material now in the hands of Lt. G. R. Bacon, U. S. A., and from

personal observations. Camp Harney (the central point of my investigations) is located on the southern slope of one of the western spurs of the
Blue Mountains of Oregon, at an altitude of about 4800 feet in 43° 80'
latitude, and 41° 25' longitude, west of Washington.

To the north of the post the country is mountainous and generally
well-timbered with forests of pine, spruce and fir, and groves of aspens
and junipers; in all other directions it is almost destitute of trees of any
size, a few straggling juniper and mountain mahogany bushes being
sparingly distributed over the different mountain ranges. The highest
and most important of these is Steen's [sic] Mountain, about seventy
miles to the south of the post, portions of which range are covered with
snow the year round.

Excepting a few warm and fertile river valleys, nearly the whole
extent of country is unfit for agriculture. About two-thirds of it is covered with sagebrush and greasewood wastes, volcanic table-lands, etc.,
the balance with nutritious grasses, and well adapted for stock-raising
purposes. As a general thing, the country may be called well watered
throughout; a continuous chain of shallow lakes extends from here to
the southwest for more than two hundred miles, and some of these are
from ten to twenty miles wide and thirty to fifty miles in length. The
water in most of them is brackish, and barely fit to drink. Fine springs,
both hot and cold, are also numerous. The many lakes form a natural
highway and convenient resting places for the immense hordes of water
fowl passing through here during the spring and fall migrations; they
also furnish safe and undisturbed breeding resorts for many species.

The climate, generally speaking, may be called mild. In the higher mountain valleys it is almost arctic, ice being formed there even in
midsummer; and many species of birds breed there which generally go
much farther north for this purpose.

Overall, Bendire's species accounts read much like a current invento-
ry of the birds found in Malheur National Wildlife Refuge (MNWR) and
the adjacent areas, describing many of the characteristic marsh, sagebrush,
and forest birds we still encounter there today. But he also mentions a few

surprises—species that have either disappeared from the area or whose occurrence in Oregon has yet to be properly documented. The former include
his accounts of the Sharp-tailed Grouse and the Least Bittern, which read
as follows:

> 126. *Pedioecetes columbianus* (Baird). Sharp-tailed Grouse.
> Only a moderately common resident, apparently irregularly distribut-
> ed. In the winter I have seen packs of from one to two hundred in the vi-
> cinity of Fort Lapwai, Idaho. They frequently roost on the willow bush-
> es along the streams, and I have seen them alight on pine trees on the
> outskirts of the timber. In the vicinity of Camp Harney they are mostly
> found in the juniper groves during the cold weather, and the birds live
> almost exclusively on the berries of these trees. The eggs usually num-
> ber from eleven to fourteen.
>
> 148. *Ardetta exilis* (Gray). Least Bittern.
> Apparently rare. I have seen it on but two occasions. It is, however, eas-
> ily overlooked, and may be rather common.

George Willett surveyed Malheur Lake in 1918 and found that the
Least Bittern "breeds rather commonly in tules on some parts of Malheur
Lake, mostly well out toward open water. Young flying by middle of July."[6]
That habitat has now been destroyed, primarily by carp.

Bendire mentions two species for which there are no current records
in Oregon—Black Rail and Baird's Sparrow. It may be questionable today
whether he identified these birds correctly, but his records are interesting
nonetheless. Regarding the rail, it may be worth noting that one of the two
reports was from April 16, at which time no young rails, which are black
and can be mistaken for Black Rail, would be present at Malheur, as adult
rails typically arrive in mid-April. In Bendire's own words:

> 152. *Porzana jamaicensis* (Gml.). Little Black Rail.
> Seen on two occasions in the swamps near Malheur Lake, where it un-
> questionably breeds.
>
> 47. *Centronyx bairdii* (Baird). Baird's Bunting [Sparrow].

May 24, 1876, I took a nest and four eggs with the parent, which I identified as belonging to this species. The nest was composed externally
of old sagebrush bark and grasses, and lined with finer materials of the
same kind and a few hairs. It was partly concealed under a bunch of tall
grass, and found on the flat about five miles below Camp Harney, on
the edge of a swampy meadow. The eggs are an elongated oval in shape,
ground color a very pale green, three of the eggs marked with irregular
spots, lines and blotches of two shades of brown (light and dark, and a
few lavender spots. The fourth is blotched throughout with a pale pink-
ish brown. In the first three eggs the markings are principally about the
larger end. Size, .72 X -55, .74 X -56, .71 X .54 and .74 X -54.

Many of Bendire's accounts of the common species he encountered are
rather brief, while others go into more detail. These accounts by one of the
first outside visitors to pay detailed attention to the birds of the region give
us a point of departure for much of what came next.

As in Arizona, Bendire was not able to collect birds and eggs without
consideration of his official duties. In 1877, he took part in the Nez Perce
War and joined the pursuit of the Nez Perce toward Canada, an assignment
in which he met Lieutenant C. E. S. Wood, later a famous Oregon writer
and transcriber of Chief Joseph's even more famous surrender speech.[7]

Later in life Spencer Baird asked Bendire if he would accept the position of Honorary Curator of the Department of Oölogy of the US National
Museum. He continued to hold this position after retiring from the army,
and in 1892 published his ground-breaking two-volume work, *Life Histories
of North American Birds, with Special Reference to Their Breeding Habits and
Eggs*. Bendire died in Jacksonville, Florida, of kidney failure on February 4,
1897. He was sixty years old.[8]

CHAPTER 4

Early Statehood, 1860–1900: Travelers and Local Lists

Alan L. Contreras

Annie Alexander • A. W. Anthony • William Brewster • Henry Henshaw • O. B. Johnson • Edgar Mearns • James Cushing Merrill • Loye and Alden Miller • Robert Shufeldt • Willis Wittich

Robert Shufeldt

The first record in ornithological literature of the presence of fossil birds in Oregon came from the work of Dr. Robert Shufeldt (1850–1934) of the US Army Medical Corps, who chronicled the work of others who collected at Fossil Lake and Silver Lake in northern Lake County in the 1880s. A national expert on bird anatomy and osteology, he wrote several journal articles about early expeditions to this area:

> [fossil birds had] been collected in the same region by Professor Thomas Condon of the University of Oregon, he being the first naturalist who discovered and collected any of the remains of fossil birds in that interesting locality. Professor [Edward Drinker] Cope's chief assistant at the time was Mr. C. H. Sternberg, now known as one of the veteran fossil collectors of this country
>
> Ex-Governor Whitaker [*sic*] of Oregon was also an early collector

of fossil birds at Fossil and Silver Lakes, and it was he who first discovered the remains of the now extinct swan, named by Cope *Olor paloregonus.*[1]

Bird fossils were not the main object for Condon, or for Whiteaker (the correct spelling), so Shufeldt is generally thought of as the first gatherer, compiler, and funnel-point for such records. His publications in 1891, 1892, and 1913 continued the process of refining his own work and that of others. However, he did not collect in Oregon himself. In mid-life he re-married, to the granddaughter of John James Audubon. An excellent biographical sketch of Shufeldt can be found in Edgar Hume's *Ornithologists of the US Army Medical Corps* (1940).

As Shufeldt was concluding his work, two other collectors and fossil hunters began working in central Oregon. Loye Miller from the Museum of Vertebrate Zoology in Berkeley, California, made an expedition to the John Day fossil country in 1899, followed by Annie Alexander to the Fossil Lake region in 1901. Miller worked primarily in the Painted Hills and other nearby sites now in or near the John Day Fossil Beds National Monument. He left us a good story of this time in his autobiography, filled with horse-drawn wagons too heavy to go uphill and other such delights.[2] In addition, the University of Oregon's Museum of Natural History issued an annotated paper about the Miller expeditions based mainly on Miller's own journal and including photographs.[3]

Miller's principal findings of living birds in his Oregon work included the state's first record of American Redstart, a "female" that he collected, though described as singing, so perhaps an immature male. His reports of living birds were published in *Condor* in 1904.[4] Among fossils, he wrote a number of articles comparing the birds of Oregon's Fossil Lake with that of the La Brea tar pits and other California sites. One of his most perceptive views was that the fossil makeup in the Fossil Lake region—heavy with ducks, divers, and gulls and largely devoid of herons, cranes, and scavengers, suggested that the lake had, at the time, been shallow and perhaps ephemeral. One of the most interesting reports was of a fossil flamingo that Miller treated as out of place.

Loye Miller's son Alden became a highly regarded California ornithologist and also studied fossil birds in addition to his work with specimens and living birds. His work on Oregon fossil ornithology appeared in several journals. The American Ornithological Society annually presents the Loye and Alden Miller Lifetime Achievement Award.

Annie Alexander

For decades, Annie Alexander and her partner Louise Kellogg collected for the Museum of Vertebrate Zoology in Berkeley, California. They made collections of birds and other fossil vertebrates (mostly mammals) in the Fossil Lake region in 1901, following directly on Miller's work and also working for John C. Merriam, whose classes she took at Berkeley.[5]

Alexander spent most of her career collecting in California and Alaska, and, as heir to a large Hawaiian sugar fortune, was the primary funding source for the museum and many of its expeditions. Thus she supported paleontology and general collections that are still in use today.

Alexander, John Kirk Townsend, and other early explorers were generalists. Some had more interest in natural history, some less, but none of them were *primarily* focused on birds. They were followed in the late 1870s by Charles Emil Bendire, who was focused on natural history when free from military duties, and be-

Annie Alexander collecting near Fossil Lake, Oregon, in 1901. Photo courtesy of the University of California Museum of Paleontology.

ginning in the 1880s by a series of military explorers whose nonprofessional interests included studying natural history. We say nonprofessional because many of them were military officers whose professional assignments
brought them to the Oregon country. Many figures in US history came to
Oregon during their professional military duties— for example, Ulysses S.
Grant and Philip Sheridan, whose names are memorialized in Oregon geographic features and communities today.

Willis Wittich and J. C. Merrill in the Klamath Basin

Some major regional lists generated by military officers assigned to Oregon
in the late nineteenth century were from the Klamath Basin. In a paper published in the July 1879 issue of the *Bulletin of the Nuttall Ornithological Club*
(vol. 4, no. 3), Edgar Mearns issued a list of birds of the Fort Klamath area
based on correspondence from Lt. Willis Wittich who was stationed at the
fort.[6] A second part of the paper appeared in the October issue of the same
journal. Wittich's list published by Mearns included the "Western Green
Black-capped Flycatching Warbler," which fortunately had become only a
Golden Pileolated Warbler by the publication of the 1910 *AOU Check-List*
and a rarely mentioned subspecies of Wilson's Warbler today.

J. C. Merrill offered his own baseline list from Fort Klamath and nearby
areas, published in 1888 in three consecutive issues of *Auk*. William Brewster commented on some of the reports, particularly raptors. This list provides some idea of what was present, with 160 species between fall 1886
and summer 1887. One paragraph gives an idea of changes in local avifauna
from the 1880s until today:

> Other species not included in either paper [by Mearns/Wittich] are
> known to occur about the Fort, but I did not obtain specimens of them.
> Among these may be mentioned the Mountain and Valley Quails, rare
> as yet but said to be increasing in numbers and extending their range;
> the Pileated Woodpecker, Purple Martin, Sage Grouse, and others.
> Among the oaks on the western slope of the Cascade Mountains within
> about thirty miles, Nuttall's and the California Woodpecker are found,
> the latter in abundance.[7]

Sage Grouse do not occur in Klamath County today except for a few in the Langell Valley about 75 miles southeast of Fort Klamath. Purple Martins are not known to breed east of the Cascade summit in Oregon; the nearest today are about 75 miles west of Fort Klamath, in central Douglas County.

Nuttall's Woodpecker has only been reported in Oregon three other times, two specimens collected at Ashland in 1881, one found dead on a roadside near Trail, also in Jackson County in 1991 and a sight record near Grants Pass in 2013. See the segment on the Williamson-Newberry report of 1857 in Chapter 2 for more on this species.

Henry W. Henshaw

Although he is known for his comments on some of Bendire's reports and for his later service as chief of the Bureau of Biological Survey, Henry W. Henshaw also visited Oregon a few years after Bendire did. His experience was mainly of dusty trails and a few birds east of the Cascades, but his reporting from the Klamath and Warner regions included Great Egret colonies and the state's first record of breeding White-faced Ibis.[8] Henshaw also collected in western Oregon, and Ira Gabrielson and Stanley Jewett note, quoting from Henshaw's memoir, that this did not always go well:

> While in Oregon an amusing incident occurred by which I fell into the clutches of the law, the first and only time in my long experience as a bird collector. Being detained in Albany, Oregon, for a few days because of a flood which interfered with the operation of the stages and railroads to the south, I employed an hour's leisure in collecting a few birds on the outskirts of the town, by no means so large then as now. Fate played me a sorry trick by leading me to collect a number of curious looking Shore Larks [Horned Larks—*Eds.*] directly in front of the house of the constable, who proceeded to instill the fear of the law into my heart by a fine of ten dollars. As, however, the birds subsequently proved to be the types of a new form . . . I have always considered that I got the worth for my money.[9]

Of some note is that Henshaw considered the Black Phoebe "common" west of the Cascades in Oregon, which was not the experience of other observers at the time. However, some of his travels were in northern California, so it is possible that his published perceptions were somewhat fuzzy in their geography.

All of these men—and to this point ornithology was essentially a male world—were itinerant observers who passed through Oregon as part of their professional duties but whose time here was limited. By the late 1800s, Oregon started to have resident observers whose records built up to the point of becoming useful local lists. Among these were O. B. Johnson and Alfred Webster Anthony, whose lists from the Portland area became the basis for knowledge of birds of that area for many years.

O. B. Johnson (dates not known)

O. B. Johnson published an extensive regional list in *American Naturalist* in 1880. In his two-part article, Johnson notes that he spent "five years at East Portland . . . two years at Forest Grove . . . and the rest of the time at Salem." Thus his list of 138 species broadly reflects the birds of the northern Willamette Valley.

Although it contains some suggestions that are, at least today, erroneous (e.g., probable breeding of the Golden-crowned Sparrow), the list is fairly complete for the valley and seems largely accurate. Some statements may reflect changes in status, for example the breeding of Lark Sparrow in the northern valley and Black-billed Magpie as "common" at Forest Grove.

Of note is that Johnson reported single records of Red-shouldered Hawk and Black Phoebe. He also noted that the sound of the Evening Grosbeak was "strikingly like the call of a lost chicken," a comparison few modern observers would be capable of making.

Alfred Webster Anthony (1865–1939) and
Harold E. Anthony (1890–1970)

A mining engineer by profession, Alfred W. Anthony is best known for his natural history collecting in western Mexico and southern California, where he spent time with Edgar Mearns. He was born in upstate New York and lived in Portland as a young man, publishing bird notes from the age of nineteen. He published a list of birds from Washington County[10] at age twenty-one. This list had a few bloopers, such as including Cassin's but not Western Kingbird. In 1886 he moved to the San Diego area and was active with the San Diego Museum of Natural History but continued spending time in Oregon and writing about Oregon birds.

The 1886 bird list also included an intriguing reference to seeing Whooping Crane fly over with Sandhills "once or twice in fall," undoubtedly mistaken as to species[11] but perhaps a genuine report of some kind of white crane. A Common Crane is said to have wintered at Sauvie Island in the 1980s[12] and that species has been reported elsewhere in the West with Sandhill Cranes; Anthony's report could have been that species.

We know Anthony today mainly because his revised bird list from the Portland area appeared in Florence Merriam Bailey's *Handbook of Birds of the Western United States* (1902 and later editions). This list of 141 species was cleaned up from the earlier published list and included a couple of things of note. He considered Lewis's Woodpecker a common summer resident in the Portland area, and he noted that Black-billed Magpie could be found in small numbers along the Columbia River bottoms. Yellow-billed Cuckoo was treated as rare in wet areas along the Columbia. This 1902 list looks thorough to a modern observer and is perhaps the earliest relatively complete and accurate list for a significant region of Oregon, along with that of O. B. Johnson.

Of note is that Anthony's son Harold E. Anthony (1890–1970) published a report in 1913 on his mammal collecting in 1911–1912 in northern Malheur County, Oregon. This report, published in the *Bulletin of the American Museum of Natural History* included several bird lists for the region around Ironside, totaling 108 species, with four species of grouse, including Sharp-tailed, now extirpated. His record of Clay-colored Sparrow is probably in error, given that he did not mention the abundant Brewer's Sparrow.

CHAPTER 5

The Northwestern Ornithological Association

George A. Jobanek[1]

On December 28, 1894, twelve young men met in Portland, Oregon. Few of them were more than twenty years old. They all had enjoyed a boyhood hobby of collecting birds' eggs and nests, but now they were aspiring to something beyond this. What they aimed for was the creation of an ornithological association, a club for young collectors in the Northwest where they could share ideas and techniques, experiences, and stories on the study of birds.

The association was the idea of Arthur Pope. Pope had been born on December 26, 1876, near Trumansburg, New York. As a boy, he became interested in the natural world, in particular birds, and spent as much time as he could roaming the countryside near the family farm. When he was thirteen, his family moved from New York to Oregon, settling in Yamhill County.

In Oregon, the teenage Pope continued his interest in birds. He read some of the ornithological magazines directed to an audience of collectors, oologists, and taxidermists. When he was fourteen he contributed "Interesting notes from Oregon," which appeared in an 1891 issue of the *Oölogist*, reporting on nesting of the Ring-necked Pheasant and Steller's Jay in Yamhill County. In the *Taxidermist* that same year, in the note "The Mongolian

Pheasant," he again discussed the nesting of the pheasant in Yamhill County, adding that hunting pressure had reduced the pheasant population.

In 1893, at the age of sixteen, he published two more articles in the *Oölogist*. In "The Sooty Grouse," he listed grouse egg sets he had collected in 1892 at McMinnville, Yamhill County. He also discussed habits of the grouse and included egg measurements. In "Notes from Yamhill County," Pope's experience as a naturalist was clear—he described collecting eggs of Ring-necked Pheasant, Ruffed Grouse, Black-capped Chickadee, Swainson's Thrush, Spotted Towhee, Dark-eyed Junco, and Lesser Goldfinch. The next year he wrote a short note on an unusual nesting of the Sooty Grouse for the journal *Nidiologist*.

In September 1894, the Naturalist Publishing Company of Oregon City, owned by G. B. Cheney, began publishing the *Oregon Naturalist*. Local naturalists now had a convenient outlet for their own notes and articles. Cheney apparently solicited them as contributors to his new journal. In the first issue, Arthur Pope published a short article on the nesting of the Steller's Jay at McMinnville, Yamhill County, which repeated some of the information in his 1891 *Oölogist* article.

Oregon Naturalist offered a means to organize bird students throughout the state and provided a forum for communication. Although the idea of a regional association probably occurred to Arthur Pope before Cheney's *Oregon Naturalist* appeared (the Cooper Ornithological Club was formed in 1893 and the reports of its secretary, Chester Barlow, ran in the *Nidiologist*, to which Pope apparently subscribed), Pope was quick to take advantage of the opportunity the new journal presented. In the first number of the *Oregon Naturalist* appeared an announcement of a possible "Northwestern Ornithologists' Association." Pope was then seventeen years old.

The November–December 1894 issue of the *Oregon Naturalist* carried the further announcement that an organizational meeting was to be held December 28 and 29 that year in Portland. That December, as planned, twelve young men met in Portland at the home of J. P. Finley to organize their new association. Arthur Pope, who had celebrated his eighteenth birthday just two days earlier, presided over the meeting. Others in attendance, in alphabetical order, were A. B. Averill, Herman T. Bohlman, A. J. Brazee, G. B. Cheney, William L. Finley, Ellis F. Hadley, Hervey M. Hoskins, W. B. Malleis, Guy Stryker, S. Rey Stryker, and D. Franklin Weeks.

This was a productive group of young men. In about five years, Pope published seventeen articles, mostly of his observations in Yamhill County. A. B. Averill became the editor of the *Oregon Naturalist* after Cheney in 1895, and published articles and editorial comments. Bohlman, 22 at the time of the meeting, was just beginning his work in nature photography. Like Pope, Hadley, of Dayton, Yamhill County, had published in the *Oölogist* before the *Oregon Naturalist* came into being, and later published many notes in that journal.

Hoskins also lived in Yamhill County, had been active since 1890, and like his peers, published notes in the *Oregon Naturalist*. At the time of the meeting he was only a few weeks shy of his sixteenth birthday. The Stryker brothers, Guy and Rey, lived in Milwaukie, Clackamas County, and collected birds and eggs, publishing their notes. In the field they were often accompanied by William L. Finley (see Chapter 6). Finley, 18 at the time of the meeting, published his first article, of what would later be more than a hundred, in the *Oregon Naturalist* in 1895.

Establishment of the Northwestern Ornithological Association

The December meeting culminated in the establishment of a new society. The young men's reasons for joining together were noble ones. "The study of ornithology being a foremost science of the day, calculated to cultivate the better qualities of man and to strengthen the powers of systematic investigation and close observation, the undersigned agree to form an association" to be known as the Northwestern Ornithological Association. Arthur Pope, the young man largely responsible for bringing these young collectors and oologists together, was elected the organization's first president. William L. Finley was first vice-president; G. B. Cheney, second vice-president; D. Franklin Weeks, secretary; and A. B. Averill, treasurer.

At this meeting, the founding members named Alfred W. Anthony, the prominent ornithologist living then in Portland, as an honorary member.

They also extended membership to Fred Andrus, a twenty-one-year-old oologist from Elkton, Douglas County, who had published in the *Oölogist* and the *Nidiologist*, and to Robert Haines, a collector from Baker City, Baker County. Frederick L. Washburn, a professor of zoology at the Oregon Agricultural College in Corvallis, also became a member.

The new members had several ambitions for their new society. "The object of this association," they wrote in their constitution, "shall be, by the active cooperation of its members, to advance the science of ornithology in all its forms, to disseminate ornithological knowledge in the Northwest, to awaken an interest in ornithology in both old and young, and to impart mutual benefit to its members." Ornithological knowledge would be disseminated through the *Oregon Naturalist*, made the official organ of the Northwestern Ornithological Association. Meetings would be annual events.

As president, Pope began work on what he saw as the principal project of the new association, the forming of a complete list of the birds of Oregon. What ornithological information on Oregon existed was in disparate sources, from the brief mentions of birds in Lewis and Clark's report of their expedition, John Kirk Townsend's narrative of his two years in Oregon with Nathaniel Wyeth, the zoological reports of James G. Cooper and George Suckley, John S. Newberry, Spencer F. Baird, John Cassin, and George Lawrence arising from the railroad surveys on the 1850s. In 1894, thirty-five years after statehood, there was no specific list of the birds of Oregon.

Pope also began gathering information from members on various bird species, with the intent of publishing a life history of at least one species a month. "It is expected that every member will send in reports," Pope wrote. "Do not hold back because you cannot make a lengthy article, but send any notes you may have. No matter how short." Requests went out for information on the Song Sparrow and the White-crowned Sparrow, the Vesper Sparrow and the Dark-eyed Junco, the Bushtit and the House Wren.

Throughout this first year of the young association's existence, members continued to find the *Oregon Naturalist* a convenient outlet for their reports and advertisements. A. B. Averill, treasurer of the Northwestern Ornithological Association, took over as publisher from G. B. Cheney with the February 1895 issue, changing the name back to *Oregon Naturalist* from its brief appearance as the *Naturalist*. He published notes on the Sharp-tailed Grouse and Ring-necked Pheasant. In the October 1895 issue William L. Finley published his first article, titled "Field notes," an account of collecting wren nests with Wade Pipes and Rey Stryker in Clackamas and Multnomah Counties.

Some Active Ornithologists Did Not Join

A few prominent people did *not* join the Northwestern Ornithological Association. Bernard Bretherton contributed articles and ads to the *Oregon Naturalist*, but he does not seem to have been a member. Born in England, Bretherton earned his living for many years as a collector of zoological specimens, at times working for a British Museum. "Zoological collecting," he advertised in the *Oregon Naturalist*. "That is my occupation, and if you desire to add to your collection species indigenous to the Pacific Coast, it will pay you to drop me a line." Bretherton also contributed an article, extended over several issues of the *Oregon Naturalist*, on Oregon mammals. Bird notes from his time as an Oregon lighthouse keeper have been published.[2]

Albert G. Prill arrived in Oregon from New York in 1890 and began publishing short notes in small journals on the birds of Linn County. His interests mirrored those of the Northwestern Ornithological Association's membership, but we find no record of his joining the association, and there is no record of whether he was even aware of the organization. Neither John Hooper Bowles of Washington nor his brother Charles, who worked in the Grants Pass area in Oregon, joined NOA.

The New State List

The November 1895 issue of the *Oregon Naturalist* carried an important article in regard to Oregon's ornithological history—the first part of Arthur Pope's compilation of the list of the birds of Oregon. Continued in the December and January issues, Pope's list included 252 species, mostly without annotation. Compiled from reports of members, with contributions also from C. W. Swallow, Bretherton, and George D. Peck, the list included a number of dubious records, such as Greater Prairie-Chicken, Brown Noddy, Greater Roadrunner, and Baird's Sparrow.

In introducing the list, Pope, writing for all members of the Northwestern Ornithological Association, remarked that "we hope, in criticising this list, that the readers of the *Oregon Naturalist* will bear in mind that it is the work of amateurs. Our association is young, and so also, for the most part, are the members, beginning in the great study of ornithology."

Beginners or not, Pope and his fellow members accomplished a great achievement by producing the first list of Oregon birds. Pope was aware of his list's limitations, admitting that it "probably does not contain all the species to be found in Oregon, yet we hope the publishing of it will give ornithologists a better idea of the avifauna of this region than they have heretofore been able to obtain, and it will certainly be of great benefit to members of the association." He estimated that the list was probably short by about fifty species.

Other accomplishments were the species reports which had appeared in several numbers of the *Oregon Naturalist*. However, notes from members were not received for some of the species selected for study. "It is to be regretted that so few of the members sent in notes for the monthly work adopted by the association. We have enough members to make valuable and interesting articles, if only all would send in a few notes. We cannot expect to accomplish good work unless all will co-operate and each one do what he is able."

At the second annual meeting, Pope saw the association's coming tasks as working to secure the passage of a state law permitting members to collect specimens for scientific study, and extending protection to more of the state's "useful" native birds. Another task, one that Pope realized would continue indefinitely, would be the "enlargement of our list of Oregon birds," in species added to the list and in details of distribution, status, and abundance of those already on the list.

The afternoon session of the second annual meeting was a reading of papers by Ellis Hadley, Hervey Hoskins, and Fred Andrus. A new member, C. W. Swallow, who had moved to Clatsop County from New England in 1890, spoke on the American Bittern and the Nashville Warbler as he had observed the species in the East. New members for the year besides Swallow were Darsie C. Bard of Portland, J. M. Gibson of McMinnville, and George D. Peck of Salem. The meeting ended with the election of new officers for the coming year. These were William L. Finley, president; Ellis F. Hadley, first vice-president; Guy Stryker, second vice-president; Arthur L. Pope, secretary; and Darsie C. Bard, treasurer.

Arthur Pope Moves On

Early in 1896, Arthur Pope, now nineteen years old, decided that journalism would be his chosen profession, and he began working at the *Yamhill County Reporter* in McMinnville. Perhaps this explains Pope's full-page advertisement in the February 1896 issue of the *Oregon Naturalist* that he was selling his

> entire private collection of Oregon birds' eggs amounting at catalogue rates to about $150.00. Not a large collection, but a choice one. Every set of which was collected in this State. All the choicest sets collected by myself and my collectors during the past five years are included in this sale. Now is the time for collectors to add something of real scientific value to their collections.

Pope must have sensed that his time for collecting was short. Besides devoting time to his new career, he discovered that he was battling tuberculosis as well.

The "Portland Annex" of the association began holding monthly meetings, often at the home of Finley, the new president. The association also welcomed as a new member Christian F. Pfluger (see Chapter 7 for his work on introduced birds). Bernard J. Bretherton, meanwhile, began an extended article on the birdlife of Kodiak Island, Alaska.

William L. Finley, nineteen when elected president of the Northwestern Ornithological Association, proved a capable successor to the energetic and productive Arthur Pope. Frustrated by his efforts to identify some birds by descriptions from texts, particularly to the subspecies level, Finley proposed that the association establish a study skin collection. The collection, to be the property of the association, would be accrued through member donations.

Finley was also concerned with how others perceived these young men in Oregon. He urged the members after their summer vacations to write up their observations "in systematically arranged notes and papers" and present them at the monthly meetings at his house. "This plan is expedient for as soon as we can demonstrate to the ornithological world that we are a wide awake, hard-working association of students of bird life, instead of mere mercenary egg collectors, then can we be assured of due recognition from the older and more scientific societies of the East[.] We are not as obscure and insignificant as we sometimes feel. We are being hopefully watched by many of our chief ornithologists, who are waiting to see of what stuff we are made. An extensive, untrodden field is open to us. Let us do what we can to explore it."

The third annual meeting of the Northwestern Ornithological Association occurred on December 29 and 30, 1896, at Salem, Oregon. Darsie Bard, the association's newly elected secretary, wrote that "the meeting was a success in every respect, members being present from all parts of the state. Rounding off the work of the closing year, reading and the discussion of the numerous reports and papers, and starting the work for the coming year on a solid, systematic basis was a task of such proportions that almost continuous session was required for its completion."

The association held its meetings in the auditorium of Willamette University. George D. Peck displayed his large collection of mounted birds. Peck had collected for many years in Iowa and included many eastern species in his display. "The most satisfactory feature of this beautiful exhibit," Darsie Bard wrote, "was the large series of the eastern and western varieties of the same species. Those of us who are struggling with that intricate taxonomical science of dividing and subdividing, with which the A.O.U. [American Ornithologists' Union] has burdened us, can easily appreciate the value of such a collection."

Members also brought with them some of their own egg sets and nests. The result was "one of the most complete and interesting collections of Oregon bird's eggs that has ever been gathered into one display." Besides association members, the general public invited to this portion of the meeting found the displays attractive and interesting.

President William L. Finley called the evening meeting to order by outlining the history of the association and the accomplishments since its establishment. He then reiterated his belief that a great opportunity to improve knowledge of Oregon birds and earn respect was before the society. Inspired by this vision, the association's members began work with new vigor and organization. Several committees were established. A field work committee held the responsibility of revising and completing the association's list of Oregon's birds that Pope had published a year ago. It also would direct the study of migration in conjunction with other ornithological groups, and it would "organize collecting exhibitions" among its members.

A library and museum committee, chaired by Herman T. Bohlman, would accept donations of specimens and books and work toward establishing the museum that Finley had earlier discussed with members through the *Oregon Naturalist*. The membership committee, chaired by Arthur Pope, was "especially instructed to employ due discretion and select only active conscientious workers." Pope was also selected to act as editor of the papers published under the association's name. A special committee, made up of C. F. Pfluger and Finley, was given the task of eradicating the House Sparrow from Oregon. There were thought to be no more than 500 sparrows in Portland, and if the committee could secure the cooperation of businesses in the city, then "their total destruction would be but a matter of a few years."

The membership voted William L. Finley to continue as president for 1897. First vice-president was Ellis Hadley; Herman T. Bohlman was elected second vice-president; Darsie C. Bard, secretary; and D. Franklin Weeks, treasurer. The meeting ended with the last of the members' papers. Darsie Bard ended his report on the meeting by noting that Arthur Pope was unable to attend because of illness. "It is hoped that before long his health may be regained."

Death of Arthur Pope and the Fading of the NOA

Arthur Pope, instrumental in establishing the association, was not to see these results of his early labors. Suffering from tuberculosis, he was forced to leave his job with the *Reporter* and return to his parents' home east of Salem; he was not to leave it again. On February 28, 1897, just twenty years old, Arthur Lamson Pope died. "He was a man faithful in all things and had left behind him an enduring reputation." Arthur Pope's grave is in the Stipp Cemetery in the town of Macleay, Oregon.

The association seemed diminished afterwards. The *Oregon Naturalist* continued under the editorship of John Martin, who had taken over from Averill with the November 1896 issue, but there seems to have been less association material. The last issue of the *Oregon Naturalist* was that of January 1898, volume 4, number 9; it no longer identified itself on the masthead as the official journal of the Northwestern Ornithological Association, as it had just the issue before. Martin tried again to publish a natural history journal, the *Petrel*, in January 1901, but it ceased after the first number.

Whether there was a formal end to the Northwestern Ornithological Association or not, members went on to other things. William Finley and Herman Bohlman continued as a team; Finley published a number of articles on birds in the early years of the 1900s, principally in the *Condor* and *Pacific Monthly*, for which Bohlman provided photographs. Both also played a part in the establishment of the John Burroughs Club in Portland and the Audubon Society which succeeded it (see Chapter 6).

Hervey Hoskins graduated from Pacific College in 1899, Haverford in 1903, and worked in McMinnville in the banking business for twenty-five years, then as county judge (commissioner) for twelve years.[3] Pfluger continued his efforts to introduce foreign birds into Oregon until his death in 1912. George D. Peck continued collecting and publishing his results in the *Oölogist*. If his report can be believed, he was perhaps the last person to see the California Condor in Oregon. He died in his 90s.

Many of the other members of the Northwestern Ornithological Association have passed on without revealing much about their lives. For some we only know them for a very brief time, when a seventeen-year-old boy had the idea of a state ornithological association and brought his friends together to enjoy their common interests. Like Arthur Pope's life, the Northwestern Ornithological Association's existence was too brief. Sadly, neither Pope nor the association have enjoyed the enduring reputation they deserve.

One question that remains is why Ira Gabrielson and Stanley Jewett, in *Birds of Oregon* (Corvallis: Oregon State College, 1940), made no mention of the Northwestern Ornithological Association in their outline of Oregon's ornithological history. In the *Birds of Oregon* bibliography, they mention only one article from the *Oregon Naturalist*, Guy Stryker's report of a Great Gray Owl at Milwaukie, Oregon.

In another paper, "Birds of the Portland Area, Oregon," *Pacific Coast Avifauna* 19 (1929): 1–54 (Jewett is first author), they refer several times to articles appearing in the *Oregon Naturalist*. Why, when *Birds of Oregon* was published 11 years later, were these articles deemed no longer worthy of mention?

As NOA faded out, the era of Finley and Bohlman began.

CHAPTER 6

The Legacy of William L. Finley and Herman T. Bohlman

Carey E. Myles

Oregon conservationist William L. Finley (1876–1953) is mostly known today for his photographs of birds taken with Herman Bohlman (1872–1943). They are remarkable photographs, sharp and well-composed, revealing not only avian life but also reflecting the exuberance and energy of the photographers and their enjoyment of the natural world. William Finley also left a rich, complex legacy for people interested in Oregon's birds through his activism and public service. However, Bohlman's work, although foundational to Finley's later career, sometimes fades into history because Finley was such a vibrant presence on the state and national scene for so long.

Finley and Bohlman's photography led directly to the establishment of several National Wildlife Refuges in Oregon and California, and Finley's continuous advocacy was key to those refuges' survival. As a founder of the Portland Audubon Society and the Oregon Department of Fish and Wildlife, Finley helped create two important Oregon entities sometimes at odds with each other today over how to best manage Oregon's wildlife.

In the 1920s and 1930s, when he had become nationally known as a popular naturalist, lecturer, and filmmaker, Finley, with his wife, Irene, and partner, Arthur Pack, made charming, funny wildlife films. They donned mountain goat or cactus costumes, featured children and family pets, and encouraged an appreciation of the natural world. However, they also used their popularity to influence public opinion on the impact of reclamation projects on wildlife refuges. In the 1930s and 1940s, Finley continued to campaign for bird habitat and clean water in Oregon and strategized with federal policy makers and officials such as Jay

William L. Finley. Photo courtesy of the Oregon Historical Society.

"Ding" Darling, Ira N. Gabrielson, and Carl Shoemaker on national legislation and funding for bird protection. He suffered a series of strokes in the 1940s, which unfortunately ended his active career.[1]

Finley was born in 1876 in Santa Clara County, California, and was from childhood captivated by birds. When Finley's family moved to Portland, Oregon, in 1887, closer to his uncle William Asa Finley, president of Corvallis College, he had a sizable collection of bird nests and eggs to bring along. He became friends with Herman T. Bohlman, who shared Finley's enthusiasm for birds and the outdoors. They spent years exploring the outdoors in and around Portland, and later farther afield, hunting to collect bird skins and eggs, honing skills in observation, and building up considerable knowledge of bird behavior.

Soon after Arthur Pope's death, Finley went to California to prepare for study at the University of California in Berkeley, and the Northwestern Ornithological Association (NOA), of which Finley had just become president, went into hiatus.[2] Former NOA members were ready, though, when Reverend William R. Lord arrived in Portland to serve at the First

Unitarian Church and, having a strong interest in birds, proposed a similar organization, to be named for noted nature writer John Burroughs. Former NOA members came together to support the new organization, and Herman Bohlman served as treasurer.[3] The John Burroughs Club later became the Oregon Audubon Society.[4]

Finley and Bohlman had become interested in photography around 1897.[5] They began taking cameras into the field, first photographing common birds around Portland, and later spending much of spring and early summer pursuing birds in Oregon and California.[6] In one of Finley's earliest published pieces, he wrote that "camera hunting" satisfied the impulse to collect, while the photographs served as proof of the photographer's

Bird photography was not easy in 1902. Ellis F. Hadley of Dayton, Oregon (leading, at left), William L. Finley (center), and Herman Bohlman (last, at right) with their pants off wading to photograph a Red-tailed Hawk's nest in 1902. Finley is carrying rope to use for climbing. Bohlman has a small hatchet in his left hand and is pulling the string attached to the camera shutter with his right hand. A posed but informative photo, typical of Bohlman's technical skill. Photo courtesy of the Oregon Historical Society Library.

knowledge of birds and their habits. The difficulties inherent in photographing birds in the field, Finley felt, added to the value of the photographs.[7]

Finley and Bohlman decided to outdo John James Audubon and il-
lustrate as many birds as possible with photographs. While Audubon had
worked primarily in a studio with dead birds as models, Finley and Bohlman wanted to depict the life of the bird. This required getting to where the
bird lived, so they hauled cameras, plates, and tripods up trees and cliffs, on
horseback, in boats, and by bike. At the turn of the twentieth century, field
photography was challenging, requiring not only technical ability but also
bringing along heavy equipment.

Finley and Bohlman made sure to document the efforts they went to in
order to obtain their shots. Bohlman set up strings to pull the camera shutter to capture themselves in action—bicycling laden with gear, climbing
trees, portaging boats, and camping on muskrat houses.[8] Bohlman wield-
ed the camera in the beginning, his photographs illustrating Finley's description of bird habitat and behavior. The two young men complemented
one another in skills and temperament—Finley was good with words and
people, outgoing and persuasive, while Bohlman, more reticent, was skilled
with tools and materials.

Finley and Bohlman preferred when possible to immerse themselves
with their subjects or to visit frequently over an extended period. In fact,
they stalked avian targets for weeks, even years. They captured images of
hummingbirds in flight and on the nest after several years of observing local hummingbird behavior and habits.[9] Similarly, they observed Red-tailed
Hawks nesting near the Columbia River from 1898 to 1902 in order to successfully create a series of photographs documenting the development of a
brood.[10]

In 1901, Finley and Bohlman took a camping and collecting trip to the
Oregon Coast with fellow former NOA members Ellis Hadley and Ross
Nicholas.[11] They managed to get out to Three Arch Rocks near Oceanside
for the first time but were short on time and not well-prepared.[12] They returned in June 1903 for another attempt. After camping for sixteen days in
the rain, waiting for the weather to turn so they could row their boat and

gear out to the rocks they got impatient and took a chance. Later Finley wrote that their little boat was tossed around in the breakers like a toothpick, and they were driven back to shore. Fortunately, their camera gear and provisions were well-wrapped.

Herman T. Bohlman and A. W. Anthony. Photo courtesy of the Oregon Historical Society.

Back on the beach drying out, Finley and Bohlman were determined to reach Three Arch Rocks. Three days of energetic effort later, they finally managed to get themselves, their camping gear, cameras, tripods, and heavy glass plates onto the rocks. There they spent five days camped precariously on a plank wedged into a rocky ledge, dodging falling rocks, baby birds, and guano.[13]

By 1903, Finley had recognized the potential of photography as a conservation tool. During their trips to Three Arch Rocks, he and Bohlman had learned it was considered sport by some to take a boat out to the rocks in good weather to shoot birds on the rocks.[14] They realized Three Arch

Rocks was a valuable nesting area for sea birds. Fortunately, their photographs of gulls, murres, puffins, and storm petrels persuaded President Theodore Roosevelt to create the Three Arch Rock Reservation for the protection of native birds in 1907. It was the first National Wildlife Refuge west of the Mississippi, established one year before Malheur.

Finley had begun studies at the University of California in Berkeley in 1899 but as important as his time at university was his involvement with the Cooper Ornithological Club in California.[15] He spent much of his time on ornithological pursuits, though he graduated with a BA in History and Philosophy. Through the Cooper Club he became connected to developments in ornithology and conservation on a national scale. He became friends with Joseph Grinnell, a graduate student in biology at Stanford who would go on to become director of the University of California Berkeley's Museum of Vertebrate Zoology. Finley also met prominent naturalists such as C. Hart Merriam—a friend of Theodore Roosevelt, Theodore S. Palmer, and Frank Chapman.

Finley's involvement with Audubon happened at the national and local levels, more or less at the same time. In 1904 Finley represented the Cooper Club at the annual meeting of the American Ornithologists' Union (AOU) in Cambridge, Massachusetts. He gave lectures on the land birds of Oregon and California and on seabirds of the Oregon coast, illustrated with lantern slides of photographs he and Bohlman had taken.[16]

William Dutcher, the president of the National Association of Audubon Societies, was impressed with Finley's photographs and lecturing style and made Finley a field representative of National Audubon, representing the Pacific Coast region. Finley was still spending much of his time in California, but he had joined the Oregon Audubon chapter (previously the John Burroughs Club) in 1902.[17] Soon he began taking a leadership role. In early January 1903 Finley and Bohlman gave an illustrated lecture as a fundraiser for the Oregon Audubon Society.[18] Finley was also active in the successful campaign in Oregon to pass a Model Bird Law in 1903.

In the summer of 1905, Finley and Bohlman visited what is now the Lower Klamath Wildlife Refuge and the Tule Lake National Wildlife Refuge to inspect and take photographs. As a child Finley had wondered about the seasonal stream of migrating birds overhead. Having learned about the expanse of inland wetlands in Southern Oregon and Northern California, the vast numbers of birds that flocked there, and the threat plume hunting posed to the birds, Finley wanted to see for himself. They went under the auspices of the National Association of Audubon Societies, tasked with documenting species through notes and photographs, and determining conditions for birds.

They left Ashland on horseback at the end of May, carrying camping equipment, three cameras, and 700 glass plates. They had arranged to pick up a small boat to travel on rivers and lakes. At Tule Lake they found huge numbers of ducks, as well as terns, stilts, and avocets. They found more grebes at Lower Klamath Lake, along with Caspian Terns, Great Blue Herons, cormorants, gulls, and American White Pelicans.[19]

They discovered that young pelicans when approached had an unpleasant tendency to "vomit up fish" in alarm and waddle off as quickly as possible. The men created a blind by using the large umbrella they had set up in the boat to shield them from the sun, adding green canvas all around. There was not much room inside with a camera on a tripod and glass plates; the photographer couldn't fully stand up, and it was hot. Finley and Bohlman took turns spending up to eight hours at a stretch crouched inside to successfully photograph the pelicans.[20] Their efforts were instrumental in securing federal protection for important nesting and resting grounds along the Pacific Flyway, although Finley would later find himself fighting a lifelong battle to maintain those protections.

The next few years were eventful for Finley and Bohlman. In February 1906 Finley married Irene Barnhart, a former classmate from University of California who would become a valuable partner in his work and a well-known photographer and naturalist writer. Later that same year, Finley and Bohlman conducted field work in California, documenting the growth of a young wild condor. Bohlman continued photographing the condor on his own after Finley was thrown from a horse and injured. At the end of the season they brought the young bird back to Portland with them. Finley also became president of the Oregon Audubon Society, taking over from naturalist A. W. Anthony.[21] He held this position until 1926, when he was succeeded by Willard A. Eliot.

It wasn't until spring and summer of 1908 that Finley and Bohlman were able to complete their inspection of Oregon's interior wetlands by visiting Malheur Lake and the surrounding marshes.[22] They were overwhelmed by the richness of birdlife they found there, but also dismayed to find that such remote marshes had been significantly impacted by market hunting. Bohlman had a White Steam automobile he had retrofitted for camping in the Oregon outback which they shipped up the Columbia River to The Dalles. From there they drove southeast to Burns, which they used as a base for several extended trips into the marshes. They found travel difficult as the water was shallow for boating, but too muddy and deep for walking. They got lost more than once among the tules and it was hard to find a place to sleep.[23]

They discovered a Western Grebe nesting ground shortly after plume hunters had been through. Finley and Bohlman were enraged to find the bodies of dead birds with just the soft breast feathers removed.[24] Still, they kept the goal of using photographs to argue for conservation in mind. Finding a dead grebe in the water next to two downy, hungry chicks sitting in their nest, they took multiple photographs, taking care over the composition. They moved the dead bird several times to create effective shots. Later when the plates were developed, they chose the most affecting one to send to the National Association of Audubon Societies. The lecture lantern slides of the photo were colored with red to show blood.

Finley and Bohlman's photography partnership essentially ended after the 1908 trip to Malheur. Later that year Bohlman married Maud Bittleston, a friend of Irene Finley's from California who would later teach vocal music at Lewis & Clark College. Bohlman took over his father's plumbing business and settled into family life. He also became a noted landscape painter. Finley and Bohlman made one more long trip together in 1912, guiding Dallas Lore Sharp and his 11-year-old son on the epic visit that became the book *Where Rolls the Oregon* (1914), later reprinted as *An Eastern Natural-*

ist in the West (2001), published by Worth Mathewson, who also produced a book on Finley and Bohlman.

Fortunately for Finley, his own family life included a robust partnership with his wife, Irene. The Finleys continued to take photographs, wrote several books together, and began making short wildlife films. William Finley became Oregon's state game warden in 1911 and took on ambitious projects, including coordination of a collection survey of the state in cooperation with the Bureau of Biological Survey. He began working toward more comprehensive goals in conservation, but his drive to protect birds continued. President Theodore Roosevelt had designated Lower Klamath Lake and Malheur Lake as National Wildlife Refuges in 1908. As game warden, Finley was most concerned with market hunting of birds on the refuges, especially for the feather trade.

Finley asked the Game Commissioners for money to hire more deputy wardens and to buy boats and motorcycles to aid in patrolling.[25] He was frustrated by the reluctance of local law enforcement and judiciary to prosecute those who broke game laws.[26] The existential threat to the refuges themselves had not yet become clear.

The size of the area set aside as bird habitat by President Roosevelt's order had seemed substantial, but Roosevelt intended multiple uses of public lands in the area, including drainage and development. Roosevelt believed rivers in the arid West should be utilized, while areas poorly suited for agriculture could be set aside for birds. In the Klamath Basin, the Klamath Project included plans to dewater Lower Klamath Lake through diversion of river water feeding the lake; the Klamath Project had been approved by Congress in 1905, three years before Roosevelt's refuge designation.

Roosevelt's apparent assumption that scientific engineering could overcome the challenges of competing water use was misguided. Crucially, the reservation lakes were not guaranteed a water supply. In addition, politically the new refuge system was weak. The refuges were to be administered by the Bureau of Biological Survey, a relatively small agency, and Congress declined to give it funding for the purpose. The more powerful Reclamation Service pushed aggressively to convert wetlands for farming and grazing.

Irene Barnhart Finley (1880–1959) carrying a box of glass photographic plates in the field. She became a well-known writer about the natural world. Photo by William L. Finley, courtesy of the Oregon State University collections.

A 1909 report on the Lower Klamath Basin commissioned by the Reclamation Service indicated farming would not succeed in the area, warned of peat fires, and suggested limited grazing was the only appropriate development. The project proceeded anyway. Malheur Lake was similarly threatened. Only the lakes were given protected status. Diversion of water for reclamation projects shrank the size of the lakes, which the Reclamation Service then used to justify decreasing the size of the refuges.[27]

Finley continued campaigning against reclamation as arriving birds crowded the shrinking amount of water, resulting in die-offs. He was bitter over the wildlife losses and infuriated by the waste, viewing reclamation projects in Klamath and Malheur as fraud perpetuated by government agencies in collusion with banks and speculators. He recognized he was fighting the dreams of hopeful homesteaders, but he believed they were being conned.

As game warden, Finley tried to discourage encroachment on the bird refuges by bolstering enforcement of existing laws against hunting, squatting, and illegal grazing. He also tried to sway the court of popular opinion, making his case in print in newspapers statewide. In 1917 the Reclamation Service cut off the Klamath River from Lower Klamath Lake. Finley was furious. In 1919 his stubborn advocacy for the bird refuges cost him his job with the state, although his frequent travel for photography, filmmaking, and lecture tours may have been another factor.[28]

The Finleys spent the next two decades writing, traveling, and making entertaining educational films with titles such as *Chumming with Chipmunks*. More sober in tone, *The Passing of the Marshlands* criticized federal reclamation projects by contrasting early film footage of abundant waterfowl at Lower Klamath Lake, Clear Lake, and Malheur Lake with later scenes of empty lake beds, dry cracked ground, and burning peat.[29]

In addition to his writing and filmmaking, Finley was a founding member of the Portland chapter of the Izaak Walton League, and served as a vice-president for the national Izaak Walton League as well as the National Wildlife Foundation. He was awarded an honorary Doctorate of Science by Oregon State College (now Oregon State University) in 1931. He also served on the Advisory Board for the Migratory Bird Treaty Act.

By this time, it was becoming clear that systematic drainage was endangering birds across the country, in many areas without providing substantial agricultural benefit. In 1934 Franklin Delano Roosevelt appointed biologist Aldo Leopold, journalist Thomas Beck, and cartoonist and conservationist Jay "Ding" Darling, a friend of Finley's, to an investigative Committee on Wildlife Restoration.

Their report highlighted the problem of having different federal agencies trying to implement incompatible goals on the same areas and recommended the government spend $25 million on mitigation efforts. Roosevelt thought he could find $1 million.[30] However, he also put Darling in charge of the Bureau of Biological Survey. Through the 1934 Duck Stamp Act and some assistance from Senator Peter Norbeck, Darling was able to get some funds to put toward restoration of the refuges along the Pacific Flyway.[31]

With a limited amount of money available, Finley and Darling decided that they had a better chance of restoration at Malheur because of the complexity of the situation in the Klamath Basin. In September 1934 they were able to arrange the purchase of 65,000 acres including water rights along the Blitzen River for the Malheur Refuge.[32] Finley supported a limited grazing program on refuge land, recognizing the necessity of local support for the project.[33] The restoration project at Malheur National Wildlife Refuge was the first large restoration project in the National Wildlife system.

Although winding down his active career, Finley continued to lend his reputation and his pen to conservation causes, including efforts to clean the Willamette River.[34] Trying to save the inland wetland bird nesting and resting grounds, he had become acutely aware of the importance of water to wildlife. In 1946, a few months before Finley's seventieth birthday, he and Irene were honored with a banquet by the Izaak Walton League at the Benson Hotel in Portland. He then retired from conservation work. Many of the well-wishers who wrote to wish him a happy seventieth birthday also thanked him for his fighting spirit. However, today it is William Finley's ability to convey the joy he felt engaging with wildlife that speaks to us most strongly across the years. Embodied in his writing, films, and his work with Herman Bohlman, this is, in the end, his most lasting legacy.

CHAPTER 7

Introduced Birds in Oregon

George A. Jobanek

Oregon had a very colorful attempt to stuff the state with European song-birds in the late nineteenth and early twentieth centuries.[1] In the latter half of the nineteenth century, acclimatization societies actively introduced birds and mammals from one region of the world to another. Recent European immigrants to America, inspired by pangs of nostalgia, tried to establish birds they knew in the Old World. This brief history of the introduction of foreign songbirds into Oregon is largely the story of one such very active acclimatization organization—the Portland Song Bird Club.

To many immigrants in the nineteenth century, Oregon was the land of opportunity. German immigrants in particular, settling in Portland, became an influential part of the city's society. The immigrants could not forget their homeland. This homesickness was manifested in a longing for familiar birds. They wished to again hear the Nightingale sing from the woods or see the Blackbird on the village green. Oregon, they felt, did not have the natural resources, and in particular the birds, of their birthplace.

By contrast to Germany, "America," the German citizens of Portland confided to the ornithologist Alfred Webster Anthony in 1890, "is a country where the flowers have no scent and the birds no song." Longing to hear again the Nightingale, they somehow missed the lovely song of the Swainson's Thrush. They wished for the Eurasian Blackbird but overlooked the similar American Robin.

On June 2, 1888, after a fund-raising campaign, the Germans of Portland formed the Society for the Introduction of German Singing Birds into Oregon, or the Portland Song Bird Club.[2] In 1888, the club contracted with a resident of Germany's Harz Mountains to capture and ship to Portland 1,000 German birds. Upon their arrival in late May 1889, the birds were for a time exhibited at the new exposition building. After the exhibition, club members released the birds.

This first liberation, the club's Secretary C. F. Pfluger told the ornithologist A. W. Anthony, consisted of nine pairs of Blackcaps (an old world warbler), sixteen pairs of Eurasian Blackbirds, eight pairs of Song Thrushes, forty pairs of European Goldfinches, forty pairs of European Greenfinches, thirty-five pairs of Common Chaffinches, thirty-six pairs of Eurasian Linnets, twenty-one pairs of Parrot Crossbills, twenty pairs of European Starlings, eighteen pairs of Eurasian Skylarks, and five pairs of European Quail. Of nineteen Eurasian Bullfinches released, sixteen were males; most of the females had died in transit. Only one pair of European Nightingales survived the transatlantic journey to escape into the Oregon underbrush. Pfluger listed forty pairs of Eurasian Siskins as included in this shipment, as were ten pairs of Wood Larks. The club liberated only a small number of European Robins, most of those received having died after arrival in Portland.

The Song Bird Club liberated most of the birds in the countryside outlying Portland and at the city park. They released the quail and six pairs of Skylarks near Salem, some Skylarks at Molalla, and Skylarks and Starlings at McMinnville.

Inspired by the efforts of the Portland Song Bird Club, perhaps even working in conjunction with it, other Portland groups sought to establish favorite species. The Oregon Alpine Club introduced Northern Mockingbirds, Bobolinks, and Northern Cardinals, birds native to other regions of the United States. The Songbird Club's Frank Dekum told A. W. Anthony that "a prominent Chinese merchant had ordered a number of song birds from his native country, as a personal contribution to the list of Oregon's songsters." His first shipment was not successful, since there was no one on board ship to care for the birds. The merchant hoped to have a second group reach Portland in time for release in the spring of 1891.

Agriculturists and sportsmen also planned importations. The Oregon Board of Horticulture instructed fruit commissioner H. E. Dosch to arrange for the importation of Great Tits from Germany. The board hoped that the birds would destroy insect pests, in particular the codling moth, the bane of apple growers. The Great Tit was never introduced, however. The Bureau of Biological Survey discouraged importations because the tit, despite being touted as the savior of the orchardist, had damaged fruit in England.

The state game warden, L. P. W. Quimby, proposed in 1903 to again introduce the Bobolink. "The bobolink ... will make a most welcome addition to our category of song birds, and it shall be my purpose, if I continue in office, to introduce this bird in this state. There are few songsters capable of pouring forth more acceptable melody than is the bobolink."[3] Quimby also suggested introducing American Woodcocks. In 1892, the Portland Song Bird Club imported another shipment of birds from Germany.

Reports after the club's introductions were immediately optimistic. All of the species supposedly survived through their first summer in Oregon and "did not wander far from Portland during the winter." Many of the imported songbirds returned the following summer to those places where the club had released them. While a few species, such as the Nightingales, Blackcaps, and Siskins, were unsuccessful, most of the others fared well. The Song Thrush, Pfluger noted, had increased "remarkably well." Northern Mockingbirds nested the season after their release at McMinnville, and Northern Cardinals were still in Portland in 1902. The European Goldfinch had "become very plentiful throughout the State, and can be seen quite often on the east side of [Portland]." Frederick Stuhr discovered four or five pairs nesting on Seventh Street.

The European Starling, too, seemed to quickly gain a foothold. Anthony, a year after their initial introduction, reported starlings seen near McMinnville. Charles Emil Bendire wrote in 1895 that the Starling had disappeared from Portland, but Pfluger observed that like the Song Thrush it had increased "remarkably well." S. H. Greene noticed it nesting about the high school building in Portland. William R. Lord also found the starling nesting in the city. Anthony listed it for Portland, but considered it rare.

The Eurasian Sky Lark apparently enjoyed the greatest success. Pfluger noted that "they have increased wonderfully since their introduction, and can be heard and seen in the proper seasons of the year upon most all the meadows, marshy and bottom lands in Oregon." Anthony noted it at Salem. Greene reported that "hundreds of them are seen in the fields and meadows in and about East Portland, and their sweet songs are a source of delight to all of us. About Rooster Rock . . . great numbers are to be seen. In fact the whole Willamette Valley from Portland to Roseburg is full of them."

Again, Bendire in 1895 reported that Skylarks had disappeared from Portland. Though the colony near Portland was reported to him to be in "a flourishing condition," he found none in May 1894. However, Lord still recorded it in east Portland in 1901. Anthony found it there yet as well. He described it as "common in open fields on the east side of the [Willamette] river; not seen elsewhere."

The Oregon Legislature even sought to protect these wonderful new residents of Oregon in the General Laws of 1895:

> Every person who shall within the State of Oregon after the passage of this act for any purpose injure, take, kill or destroy or have in his possession, except for breeding purposes, sell or offer for sale any nightingale, skylark, black thrush, gray singing thrush, linnet, goldfinch, greenfinch, chaffinch, bullfinch, red-breasted European robin, black starling, grossbeak, Oregon robin or meadow lark or mocking bird, shall be guilty of a misdemeanor.

The nests and eggs of these birds were also protected.

Other scientists joined Anthony in raising objections and criticisms as they became aware of the activities of the Portland Song Bird Club. C. Hart Merriam and T. S. Palmer, of the Bureau of Biological Survey, recognized the danger "of discovering another new pest as the English [House] Sparrow." Palmer urged in 1899 that introductions be restricted by law and carefully controlled. He considered many of the species liberated by the Portland club as "of doubtful value and likely to prove injurious."

A few years later in 1908, when the club was making its last introductions, Joseph Grinnell, editor of the *Condor* and the preeminent ornithologist of the West Coast, called their actions "idiotic." "The next thing we know we will have Chaffinches and Goldfinches to deal with along with the 'English Sparrow problem.' The Audubon societies should bend their efforts against the introduction of foreign birds, if they wish to keep our native avifauna intact."

Grinnell was not supported by the Oregon Audubon Society of Portland. Rather than opposing the introductions, the society announced plans only a few years after Grinnell's appeal to again attempt to introduce Northern Cardinals into Oregon. "We are convinced they will thrive well here," Emma J. Welty wrote in *Bird-Lore* (1912). The society was unable to obtain birds, however. William L. Finley, president of the Oregon Audubon Society, seemed reluctant to enact the plan and perhaps dissuaded the organization from pursuing its attempt further.[4] Northern Cardinals seen by Stanley Jewett in Douglas County in 1930 were escaped cage birds.

The House Sparrow problem Grinnell alluded to was the perfect example of the implicit hazard in introducing foreign birds into a native avifauna. Brought from Europe to the eastern United States in the 1850s as a means of ridding cities of insect pests and as a nostalgic reminder of the European homeland, the House Sparrow soon began a remarkable spread across the continent. In Oregon, the little sparrow's rise to abundance was contemporaneous with the Song Bird Club's attempts at introductions and should have been an obvious warning of the danger of their actions. It is not clear when or how the House Sparrow first arrived in Oregon; Ralph Hoffmann speculated that it came with freight trains from the East.

Certainly starlings and Skylarks, and perhaps European Goldfinches, seemed to adapt to Oregon. Yet by 1929 Stanley Jewett and Ira Gabrielson could write that all the foreign birds had disappeared. The European Starling disappeared for an unknown reason. Gabrielson and Jewett (1940) noted it was gone by 1901 or 1902. It ultimately reached Oregon from an introduction a continent's width away as a result of its aggressiveness and fecundity, which the birds released in Oregon apparently lacked. Eurasian Skylarks had also vanished from the fields of the Willamette Valley. Their "abundance" might have been an illusion, Horned Larks or American Pipits or sparrows misidentified. Hundreds had been released for naught.

The club's efforts now seem amusing, born of a curious chauvinism, of a discomforting nostalgia quieted only by the singing Nightingale. Most likely the members stopped their importations when they realized the futility of their efforts, or ran out of money, or exhausted their energies.

Perhaps a club member followed Anthony's advice and visited a wooded ravine in May. The song of a Swainson's Thrush, or a Black-headed Grosbeak, or a Pacific Wren, echoing melodically from a tangled thicket, would surely have caused the listener to realize that the club's efforts had not been necessary at all: Oregon was not a place where the birds had no song.

There are seven introduced bird species which successfully maintain populations in at least part of Oregon without assistance. The status of these seven—the Gray Partridge, Chukar, Ring-necked Pheasant, Wild Turkey, Rock Pigeon, European Starling, and House Sparrow—is discussed in Marshall et al., *Birds of Oregon* (2003) and in Joe Evanich's 1986 overview in *Oregon Birds*.[5] The pheasant in particular was widely noted at the time, and Oregon Agricultural College (later OSU) professor William T. Shaw, whose taxidermy bird mounts won a gold medal at the Lewis and Clark Exposition in Portland, Oregon, in 1905, published a book titled *The China or Denny Pheasant in Oregon* in 1908. These pheasants were called "Denny Pheasants" for a time because of how they got to Oregon. In his history of the Oregon Fish and Game Commission through 1938, Frank Wire described this bird introduction.

> In 1880 and 1882, Judge O. N. Denny, then serving as Consular [*sic*] General in Shanghai, conceived the idea that the Willamette Valley, being similar in climatic conditions to that of China, might be a suitable place for Chinese pheasants. He sent fifty of these birds to be released on his old home place at Peterson Butte, a few miles from Albany. They were protected for a period of ten years, after which an open season of

six weeks was permitted, which was later lengthened to two months. These birds increased very rapidly and a report states that in 1893 more than 30,000 were killed in one county.

Gene M. Simpson saw the possibility of rearing Chinese pheasants in captivity and in 1900 established five miles north of Corvallis the first private game farm in the United States. He had been successful in his efforts so the Board of Fish and Game Commissioners in 1911 leased his farm for a period of three years and employed Mr. Simpson to raise pheasants for the state. During the fall of 1911, 1206 pheasants were reared and released in 1912. These birds were distributed in such sections where pheasants were not plentiful. During this time the Commission also purchased 120 pairs of Hungarian Partridge for liberation.[6]

Denny's brother John had received the pheasants and assisted in their propagation.

A few other introductions should be mentioned. White-tailed Ptarmigan and Northern Bobwhite were introduced and remained semi-viable for many decades (over a hundred patchy years in the case of the Bobwhite). It is possible that the ptarmigan remains in the Wallowa Mountains, though it is likely extirpated. The Monk Parakeet, an escapee, rattled around the northern part of Portland for decades and finally seemed gone in the early 2000s. The Mute Swan has become controversial because of its essentially harmful nature. A few can be found here and there in the state.

The Trumpeter Swan and California Quail, species which were native only to a small portion of the state, have had their Oregon ranges expanded due to transplanting. Sharp-tailed Grouse, probably extirpated in the 1960s, have been reintroduced in Wallowa County but may not remain viable there.

CHAPTER 8

Government Surveys and Academic Study, 1901–1939

Alan L. Contreras, George A. Jobanek, and Noah K. Strycker

Vernon Bailey • Johnson A. Neff • Morton E. Peck • Alfred C. Shelton • George Miksch Sutton • Edward Preble • A. Roy Woodcock

As ornithology became a free-standing field, Oregon saw two principal forms of professional study develop. One of these was organized field study by government agencies, mainly those of the federal government. The other was the emergence of academic study of birds by Oregon college faculty and students. This book includes material on these advances up to about 1950; see *As the Condor Soars* for information on modern research, conservation, and agency work.

SURVEYS BY US GOVERNMENT AGENCIES

From the mid-1800s through the middle of the twentieth century, the US government sponsored significant field work by employees and contracted workers. The Pacific Railroad Reports and the work of Charles Emil Bendire, Willis Wittich, and others contained the results of work under frontier conditions in the 1800s. By the end of that period the frontier was, if not gone, at least pushed back into pockets within tangible patterns of civilization.

What we think of today as the US Fish and Wildlife Service had its beginnings in the Division of Entomology of the US Department of Agriculture. Ornithological study as a named function of a unit of the US government began, in the legal sense, on March 3, 1885, when the division added a new clause to its mission statement set by Congress that gave it the new duty of "economic ornithology, or the study of the interrelation of birds and agriculture, an investigation of the food habits and migration of birds in relation to both insects and plants."[1]

In 1886 the unit to do this work was designated the Division of Economic Ornithology and Mammalogy. In 1896 it was relabeled the Division of Biological Survey under the leadership of C. Hart Merriam, brother of ornithologist Florence Merriam Bailey. It became the Bureau of Biological Survey (hereinafter the Bureau) in 1905 and used that name until 1940, when it was combined with the Bureau of Fisheries and renamed the Fish and Wildlife Service, both having been moved to the Department of the Interior in 1939. Many of the early ornithological explorations of Oregon occurred during this roughly fifty-year period and involved federal field workers.

In this chapter we discuss several of the more prominent field collectors and observers. In general these are the people who published free-standing books or major articles about Oregon birds. Many more were active in Oregon, particularly collectors working in the first twenty years of the twentieth century. A longer discussion of who they were and where they worked can be found in the introductory material in *Birds of Oregon* (1940).[2]

Vernon Bailey

Vernon Bailey, husband of ornithologist Florence Merriam Bailey (see Chapter 9 for more on F. M. Bailey), was primarily a mammalogist and is best known for inventing a line of live-traps for mammals and for his 1936 book *Mammals and Life Zones of Oregon*, published by the Bureau. The book contained a number of bird lists as well as including Native American names for many of the species it discussed. Bailey was a rather patchy ornithologist compared to his wife; witness this description from a 1909 survey in Coos County:

> *Aphriza virgata*. Hundreds of snowy little snipe were running out and in with the waves on the beach as we came down from Umpqua Bay to Coos, Oct. 9. None were taken but I am sure they were Surfbirds in the light winter dress.[3]

Range Bayer, in whose collection of Bailey's notes this description appears, notes that "his description better fits Sanderlings." Not really a close call by 1909, one would think, as Surfbirds were known to use rocky outcrops and not dash around on beaches in flocks. The behavior of Sanderlings was even described in F. M. Bailey's guide, published several years earlier. Yes, the word "snipes" was used casually as a term for small shorebirds, but Bailey was an experienced biologist.

Too often as we research the baseline data for Oregon ornithology, we come to Ira Gabrielson and Stanley Jewett's *Birds of Oregon* (1940) and find that some of the potentially most useful and detailed material exists only in field notes, not in published sources. Edward A. Preble's field notes from Malheur County are a good example of that data. Fortunately, we have Noah Strycker's modern outline of the trip and annotated transcription of the notes available. This serves as a sample of what field work was like in the early twentieth century.

Edward Preble

Malheur County lies in far eastern Oregon and includes the southeastern corner of the state, where Oregon's ecology resembles that of Nevada more than that of the rest of the Oregon high desert; in addition to some unusual birds, there are cacti and kit foxes here—not something one often sees elsewhere in the state. Although Edward Preble did not find the Gray-headed Juncos that are now known to breed in the Trout Creek range or the Virginia's Warblers that now occur on occasion, his early trip found the first state records of White-throated Swift and added a great deal of information about the state's ornithology that can't be found anywhere else.

Preble's field notes serve as an example of the kind of route that early field operatives covered in their work. They document his trip from June 1 to August 12, 1915, starting near McDermitt, Nevada, just south of the Malheur County line, and ending at Ontario in northeastern Malheur County. It is clear that he had use of a vehicle during this expedition, but detailed logistical information about the trip is lacking. Preble recorded observations at McDermitt from June 1–9. He specifically mentions in the species accounts a canyon eight miles northeast of McDermitt which is in an aspen grove quite distinct from the valley floor at McDermitt.

He then headed westward to nearby Disaster Peak, spending June 9–17 on the east side of the mountain from 5,600 to 8,200 feet and then traveling June 17–19 from Disaster Peak northward to the Owyhee River. From June 20 to June 24, he was headquartered at Rome near the Owyhee River. From Rome, he headed east to Jordan Valley, where he stayed from June 25 to July 2, and then spent July 3–11 in the Cow Lakes region. He explored around Sheaville from July 12 to July 14, Rockville from July 15 to July 19, and Watson from July 20 to July 25, then moved northwest to Cedar Mountain, where he remained from July 26 to July 29.

After leaving Cedar Mountain, he proceeded northwest to Skull Spring, where he stayed July 30–31, and to Riverside, where he recorded observations from August 1 to August 3. Leaving Riverside, Preble concluded his trip in the northeastern corner of the county, staying in Vale August 7–8, in Owyhee August 9–10, and in Ontario August 11–12. Preble's route covered much of the county, but he missed some territory in the southwest, southeast, and northwest corners.

Preble organized his field notes by location. He may have written a physiography for each of the fifteen sites he visited. However, only two remain (Disaster Peak; Disaster Peak to Owyhee River). These physiographies provide detailed descriptions of terrain and itinerary; a full set would make it possible to retrace Preble's footsteps more accurately.

Besides documenting the birds observed on his trip, Preble's field notes give a taste of the conditions under which he was working. As he traveled by day and camped by night, Preble was at the mercy of the elements, yet he

Edward Preble in 1923.

mentions no physical discomfort. In his record of sage grouse observed on Disaster Peak, it only becomes clear by implication that a snowstorm had hit on or near his camp just a few days before:

> On June 17, when we crossed the hills from 6000 to 6800 feet in altitude to the northward of our camp at the east base of Disaster Peak, we saw an aggregate of nearly 300 sage grouse. They were in small companies, sometimes with flocks of one sex and sometimes of both but all without young. Probably the broods were killed by the snowstorm of the 11th.

Preble collected a number of specimens on this trip, including parts of birds (such as stomachs), representing such species as Sora, Western Sandpiper, Common Nighthawk, Common Poorwill, Canyon Wren, Swainson's Thrush, Yellow Warbler, MacGillivray's Warbler, and Song Sparrow. Nearly sixty specimens of thirty-eight different species from Preble's Malheur County trip are currently held in the Smithsonian Institution collections (US National Museum).

Interestingly, the Smithsonian holds three well-documented Fox Sparrow specimens from Preble's trip (USNM 242871-3), although Preble does not mention any sightings of this species. His field notes relate information about species distributions and breeding activity, sprinkled with entertaining stories about bird behavior and anatomy. For instance, Preble remarked at some length about a pair of Prairie Falcons seen during his stop at Skull Spring:

> A pair of Prairie Falcons visited the ranch, near which we were camped, each day, usually appearing just after sunrise in the morning. They usually made a raid on the flocks of chickens and young turkeys, and, according to the ranchers, had carried off a number of them.

In a White-throated Swift account at Watson, Preble wrote:

> Four individuals were seen about 4 miles southwest of Watson, where the valley narrows and enters a canyon, on the evening of July 24. One of these birds was secured. In the small pouch which is formed in the throat of the bird by the distention of the elastic skin was a considerable number of small insects which it had just captured. Some of these were still alive but firmly held in place by the tongue and by the mucous which is secreted by the bird.

This was the first report of White-throated Swift in the state. Preble listed 106 species overall. Notable in his lists were Broad-tailed Hummingbird and Veery.

The principal value of Preble's early field notes is not that he found a lot of unusual birds, but that his careful statements about a bird's status provide in some cases the only reliable historic benchmark for the species' status in Oregon, a benchmark that is very helpful as we consider the status of a species today. For examples, see the accounts for Mountain Quail (now very local in the mountains of SE Oregon), Bobwhite (introduced, now extirpated), Sage Grouse (range reduced), and Eastern Kingbird (less common in far SE Oregon).

Preble's handwritten field notes were meticulous. Photo courtesy of the US Museum.

Oregon ornithology greatly benefited from the US Bureau of Biological Survey's decision to send Preble, then a forty-four-year-old assistant biologist, for the Malheur County field survey. A rugged outdoorsman and authoritative ornithologist, Preble was best known for a two thousand–mile canoe journey he took with famed naturalist Ernest Thompson Seton in 1907, which was chronicled in a popular book of the day.[4] Though Preble never attended college, he excelled as a naturalist and ended his career as an editor of the well-respected *Nature* magazine. —*Adapted and revised from Strycker,* Early Twentieth Century Ornithology in Malheur County, Oregon *(2003)*[5]

Morton E. Peck

Morton Peck, later a professor at Willamette University, may be best known for his comprehensive *A Manual of the Higher Plants of Oregon*, published by Oregon State University Press in 1961.[6] Though long out of print, the volume is still a coveted reference among botanists.

Peck's bird observations, published in *The Condor*, occurred from June 22 to July 25, 1910, in the Willow Creek Valley of northeastern Malheur County. In his paper Peck listed 74 species in all, including eight that were not noted by Anthony or Preble. His unique observations included Yellow-billed Cuckoo and Gray Catbird. It is worth noting that he listed Tricolored Blackbird while omitting the abundant Red-winged Blackbird; this was most likely a mistake.

One of his Swainson's Thrush specimens was later determined to be a Veery, one of the southernmost for Oregon in breeding habitat.[7]

—Noah K. Strycker contributed to this account

Alfred C. Shelton[8]

The southern Willamette Valley at the beginning of the twentieth century was blessed with several active ornithologists, the most well known of whom were probably Dr. Albert Gregory Prill of eastern Linn County, Arthur Roy Woodcock of Corvallis, and Alfred C. Shelton of Eugene. Shelton, in particular, made many important contributions to the ornithology of the southern Willamette Valley and of Oregon as a whole. He was one of the first college educated ornithologists (University of California, Berkeley) to work in the state. As such, he represents the beginnings of transition from the self-trained ornithologists such as Robert Ridgway and Charles Emil Bendire to the modern pattern of separation between professionally trained and amateur bird students.

In the time when collecting was the only efficient way to catalog an area's birdlife, Shelton was an indefatigable collector. To look through the University of Oregon Museum of Natural History's bird collection, almost entirely Shelton and Albert G. Prill's skins, is to marvel at the immense effort and time which went into ornithological research of Shelton's era. Prill once commented that it took him half an hour to prepare a sparrow skin and half a day to put up a goose; Shelton, after spending the day collecting numerous birds perhaps ranging from hawks to sparrows, would spend a large part of the night making study skins.

The immense effort that went into Shelton's field work is further appreciated when the conditions of the time are realized. Shelton effectively covered diverse habitats and areas of Oregon at a time when roads were crude. He writes of taking the train to Marcola, the trolley to Springfield, and the

Alfred C. Shelton at the University of Oregon about 1916. Photo courtesy of the University of Oregon archives.

stage to McKenzie Bridge. Field work often meant establishing camps for one- or two-week periods. Even nearby Spencer Butte was seven miles from Eugene; this distance did not allow many trips to town.

Although Shelton concentrated a great deal of his field work on the birds of Lane County, he also made excursions to other areas, such as Bend, Moody, Galice, and Netarts Bay. He was one of the first ornithologists to work in the Umpqua and Rogue River Valleys. Because Shelton published but one monograph, and that on land birds of what amounted to Lane County, his field work remained little known.

Gabrielson and Jewett's *Birds of Oregon* research did an inadequate job of checking bird collections, in particular Shelton's. Although Jewett in 1930 and Gabrielson and Jewett in 1940 considered Overton Dowell Jr.'s 1923 Mercer specimen as the first Green Heron collected in Oregon, Shelton had collected one in 1915. Again, despite Gabrielson and Jewett's statement that H. H. Sheldon took the first Oregon Red-eyed Vireo specimen in 1916, one month before Shelton took one at Oakridge, they overlooked the fact that Shelton had collected a breeding female in Grants Pass the year before. Indeed, Gabrielson's 1931 article, "The Birds of the Rogue River Valley, Oregon," would have been much improved had he consulted Shelton's fine collection.

Another of many minor errors in *Birds of Oregon* (1940) which hinder realization of Shelton's contribution to Oregon ornithology appears on page 542. Here it reads that

> Vernon Bailey, Jewett, and Alex Walker were among the members of a Biological Survey and Oregon State Game Commission party that collected on the Three Sisters, on several dates between July 11 and 17, 1914, the first breeding [Rosy Finches] taken in the state.

Overlooked was the fact that Shelton was a prominent member of that party—in fact, Shelton, accompanied by Jewett, climbed to the top of the Middle Sister on July 12 and collected four finches. Since *Birds of Oregon* (1940) was written entirely by Gabrielson, this omission of Shelton's contribution to that record is understandable but unfortunate.

Even though it only covered Lane County and some adjacent areas, Shelton's annotated list, *A Distributional List of the Land Birds of West Central Oregon*, issued by the University of Oregon as part of its bulletin series,[9] was the first Oregon bird list that could be considered a free-standing *scientific* publication. It did include some information from earlier publications, but it was based largely on Shelton's own collecting and observations, supplemented by material held by the Museum of the UO Department of Zoology.

Because Shelton was one of the first people to systematically approach Oregon bird study in lab and field, some additional detail about his life and work seems appropriate. He was born in 1892 in Petaluma, California, where his early interest in birds attracted local attention and allowed him to be chosen as a member of the Cooper Ornithological Club at age eighteen. He remained active as a student at the University of California at Berkeley. At the recommendation of Joseph Grinnell from the Museum of Vertebrate Zoology, Shelton was hired by the University of Oregon museum to create a zoological collection there.

Frog Camp party at the base of Three Sisters, July 1914. Left to right standing: Stan Jewett, Alex Walker, Vernon Bailey (chief field biologist, US Biological Survey), E. A. Goldman, Alfred C. Shelton (University of Oregon), Morton E. Peck (later of Willamette University), D. E. Lancefield. On the ground right: Bruce Horsfall (artist), Jack Frey and the dog, Sport. Photo by Alex Walker.

He worked for professor John Bovard (see Bovard's 1917 course book for ornithology depicted on page 81) and also with Stanley Jewett, then a state biologist and future co-author of the 1940 *Birds of Oregon*. One of these collecting adventures, focused on the red tree vole, is chronicled in an article in the *Oregon Historical Quarterly*.[10] Shelton is credited with the first Oregon specimen of a northerly subspecies of Song Sparrow, *Melospiza melodia caurina*.

Shelton's collecting duties took him throughout western Oregon, including coastal areas north to Netarts, and in early September 1916, four years after Dallas Lore Sharp's visit, he went with another observer to the Malheur area for a week. Shelton's original field notebooks are held by the

Samples from Alfred C. Shelton's museum collection. Alfred Cooper Shelton collected this towhee at the Long Tom Marsh near Elmira, Lane County, Oregon, on April 2, 1914. The specimen is listed in both his collecting catalog and the accessions file card. Special thanks to Dr. Pamela Endzweig, Anthropological Collections Director at the University of Oregon Museum of Natural and Cultural History, for assisting with access to specimens and Shelton's original journals. Photos by Tye Jeske.

University of Oregon Museum of Natural and Cultural History. In 2002, Oregon Field Ornithologists (now the Oregon Birding Association, with a somewhat different focus) issued an expanded, annotated reprint of Shelton's book, edited by Noah K. Strycker. — *George A. Jobanek*

THE GRINNELL SYSTEM OF RECORDING FIELD DATA

Bird observers take field notes in many ways. Today, many don't really take notes at all, relying instead on simple data-entry systems like eBird (see Chapter 19). Shelton was trained by Joseph Grinnell at the Museum of Vertebrate Zoology, associated with the University of California, Berkeley. Grinnell had his own particular way of maintaining field records and their associated journals and, if applicable, collection records. This became known as the Grinnell System and has become something of an icon in the community of ornithology and at the MVZ.

For a thorough outline of Joseph Grinnell's system of record-keeping, see the late Steve Herman's extraordinary walk-through of how the system works, *The Naturalist's Field Journal: A Manual of Instruction Based on a System Established by Joseph Grinnell* (Vermillion, SD: Buteo Books, 1986), or the less detailed but useful exposition by J. Van Remsen in "On Taking Field Notes," *American Birds* 31, no. 5 (1977): 946–948, readable on the SORA online database.

For a brief discussion of the question of handwritten notes vs. electronic databases, see Jennie Duberstein's blog post for the American Birding Association on March 8, 2014. The late Steve Herman commented recently[11] on changes in how people keep field records:

> It's not just the writing. It is largely that the system demands the observer know where he is, know what he's watching and when. And time must be made every day to edit the material and make sense of it when it goes from field notebook to journal. It's a demanding system. So it makes a permanent, theoretically accessible record, but it also has had a role in educating the writer. It even improves penpersonship.

University of Oregon Bird Study, 1917

An interesting connection exists between Alfred C. Shelton and one of the state's earliest ornithology courses available to the public. John Bovard, Shelton's supervisor at the UO Museum, was the faculty member of record for this course. At the time, Bovard was a professor of zoology and an active field biologist, though his professional focus was on physiology and education. The published volume of material for the course is quite sizable and detailed and has 190 images hand-glued into it. A 240-page hardbound quarto, it was issued by the university's extension division. Bovard was also a mentor to Hubert Prescott.

University of Oregon Bird Study sample page. Photo by Alan Contreras.

Hubert Prescott (1899–1988)

Hubert Prescott was born in February 1899 on a farm on the Nehalem River near Jewell, Oregon. He loved bluebirds as a child and contributed articles about bluebird houses to the *Oregon Teacher's Monthly* in 1913 and *Bird-Lore* (later *Audubon* magazine) in 1917 while living near Ashland, Oregon. He worked for the University of Oregon for a time and eventually for the US Department of Agriculture in Forest Grove, Oregon. By the time he retired in 1965, Western Bluebirds had almost completely disappeared from the northern Willamette Valley.

John Bovard (right) and Hubert Prescott collecting in coastal Lane County, May 1920. Photo courtesy of the University of Oregon archives.

In 1971 after six years of searching the area Prescott finally found a pair nesting in an old nest box on Chehalem Mountain southwest of Portland. In talking with the landowner he found there were about a dozen pairs nesting in the area, all in old nest boxes people had on their property. By 1975 he had installed over 200 nest boxes in several areas. According to a memorial description:

> He would approach each farmhouse . . . and explain his mission. He was rarely refused. He would mount nestboxes on trees, fenceposts, or sides of barns, always in suitable habitat. Befriending the

Hubert Prescott in his seventies, cleaning a bluebird house. Photo courtesy of the Hubert Prescott Bluebird Recovery Project.

> landowners, he would monitor the boxes, keep impeccable records, and photograph whichever kind of bird was using the box, presenting a copy to the landowner.

By 1975, he had built over 200 nestboxes, and erected them on Chehalem Mountain, and also on Parrett, Cooper, and Bull Mountains, and in the West Hills of Portland. He also installed nestboxes in the Molalla and Colton areas, and in the West Linn-Stafford areas. It was impossible for him to monitor this many boxes without help, so he enlisted the help of the Portland Audubon Society. . . .

The amount of territory that he covered was astounding. In the 1980s, when I thought I was setting up boxes in new territories, the landowners would tell me that an elderly gentleman had been there some years earlier and put up boxes. Even into the 1990's it wasn't unusual to come across a nestbox that Hubert had erected years earlier.

His activity also inspired Elsie Eltzroth of Corvallis, Al Prigge and Barbara Combs of Eugene, and others to establish local "bluebird trails" where multiple boxes were successful in bringing the bluebird back as an expected bird in the Willamette Valley. In addition to his work with bluebirds, he was active in Purple Martin box projects in the 1970s, particularly with Tom Lund in Lane County. Prescott died in June 1988.

The kind of work done by Prescott does not fit easily into the categories of professional/amateur, nor is it scientific in the usual sense. It is a kind of volunteer "wildlife management," done for a species that was not actively "managed" at the time but the population of which had declined drastically since the 1930s. People doing this kind of work often became the de facto experts in some aspects of the species' breeding biology. The late Elsie Eltzroth in particular published several journal articles originating from her bluebird recovery work.—*Includes material contributed by the Prescott Bluebird Recovery Project*

George Miksch Sutton (1898–1984)

George Miksch Sutton, one of the giants of American ornithology, was not an Oregon bird researcher in the usual sense, but his two childhood years in Oregon set him on the path critical to his career and artistic expression.

Sutton was born in Bethany, Nebraska. When he was about nine years old, he moved with his family to Ashland, Oregon. To reach town from his home, Sutton would cross Ashland Creek over a little bridge. There he would stop and gaze into the stream below. The remembrance of Ashland Creek, he wrote over seventy years later, became "a continuing part of me—a part that would refresh me when I needed refreshing and calm me when I needed calming. No other stream in my whole life would do this for me."

Birds had fascinated him, and the move to Oregon introduced him to unfamiliar species. His copy of Frank Chapman's *Bird-Life* was of little help in identifying them, since it dealt with eastern species, but this did not affect young Sutton's curiosity and attention. Returning from a climb of Ashland Butte with his father, he picked up dead birds stricken by an ice storm from the grounds of a cemetery. He pasted their feathers on cards and added the species names he thought appropriate.

Once he captured two male Western Tanagers, so locked in combat that they fell to the ground before his feet and he had but to throw his cap over them. Their beauty entranced Sutton and stirred uneasy feelings: "All that beauty, that exquisite beauty, I wanted to keep, to have for reference, to show to people by way of making clear how wonderful birds could be." He knew nothing of preparing bird skins, so he reluctantly let them go.

A year after moving to Ashland, the Suttons moved again, to Eugene. Sutton's father was a preacher and college instructor, and the elder Sutton now took a position at the Eugene Bible University (now Bushnell University). They lived next to the campus, not far from the University of Oregon. Here again, new birds enchanted the young Sutton. In the spring of 1908, playing with his sister in their yard, he saw a Violet-green Swallow snatch a feather from the air.

The children immediately ran to the house and raided pillows for more feathers and gathered plucked feathers from the woodshed floor. Swallows began to mill around them at once, and for a week followed the children whenever they went outside. "Seated at meals in the dining room, we perceived that the swallows were watching us as they flew back and forth just outside the window. Occasionally one tapped at the glass as if to remind us of our duty."

The 10-year-old Sutton wrote of this experience to William L. Finley, then president of the Oregon Audubon Society of Portland, and Finley answered his letter. Finley wrote a monthly column for the Audubon Department of School and Home magazine. The column on swallows which subsequently appeared described in detail the attractiveness of feathers to Violet-green Swallows, but did not quote from Sutton's letter.

When not watching swallows or other birds, young Sutton would visit the University of Oregon library. Chapman's *Bird-Life* was a frustrating reference in a western state so Sutton sought a more appropriate guide. He discovered Florence Merriam Bailey's *Handbook of Birds of the Western United States*. With this book, by carefully reading each species account, he could identify any bird he might encounter. Modern observers will recognize the descriptions, some of which were condensed from Robert Ridgway's *Manual of North American Birds* and *Birds of North and Middle America* and later reprinted in species accounts in *Birds of Oregon* by Ira Gabrielson and Stanley Jewett.

Furthermore, the introduction included a list of the birds of Portland by Alfred W. Anthony, providing a good check on the young ornithologist's more surprising identifications. Also in the introduction was a veritable how-to of ornithology: a section by Mrs. Bailey's husband, Vernon, chief field naturalist of the US Biological Survey, on collecting and preparing birds, nests, and eggs. Florence Bailey added instruction on note-taking.

More significant to young Sutton were the full-page black-and-white plates by Louis Agassiz Fuertes, an artist whose very name fascinated Sutton. Fuertes' birds had that spark of animation that would later become a Sutton mark as well. His Varied Thrush sang from the page, and young Sutton, sitting alone in a quiet library, must have heard it ringing in his head. Although he had made sketches of birds before moving to Oregon, once in Eugene Sutton began drawing birds "in earnest," directly influenced by Fuertes' plates in Bailey's *Handbook*. He drew in pencil and saved the best and pasted them together in two rolls of paper.

The names he added showed more time spent in the field than in the classroom—the Screaming [Steller's] Jay, the Least Pheode [?], the Red Wing, the Gairdener [Downy] Woodpecker, the Dwarf Chewink [Spotted Towhee], the Vaired [*sic*] Thrush. At 10 years old not as anatomically precise as he would later become, the Red-winged Blackbird was a structural monstrosity with extra leg bones. Sutton's drawings were crude, but they were a significant step on a distinguished artistic career. Years later, he would display them proudly.

Living close to the university also enabled Sutton to visit the university's bird collection, housed in Deady (now University) Hall. Professor A. R. Sweetser, a botanist, took the boy under his wing and gave him the run of the biology lab. He allowed Sutton to arrange the collection. The number of skins in 1908, a few years prior to Shelton's collecting, must have been small, for the 1909–1910 university catalog said in describing the zoological museum that "the specimens in the museum which are typical of Oregon fauna are few, and nothing would be appreciated more by the [zoology] department than the gift of skeletons or skulls of Oregon animals or the skins of Oregon birds."

Whether few or not, Sutton delighted in keeping the skins in proper sequence. "I greatly enjoyed the biology laboratory," he wrote in his autobiography. "I came to feel that I was welcome there, even needed, for I kept the bird skins in order." Ten-year-olds, however, hear seductive whispers from all manner of otherwise innocent things, and Sutton, not yet bound to the mast of maturity, fell victim to the wiles of the building's polished wooden stair railing. Unable to resist sliding from the third floor to the first with his friends, he was banished from the building by a campus security guard.

This was a major setback to the aspiring young ornithologist, but only temporarily so. "Being denied access to the bird skins at the university forced me to build up a collection of my own." His first attempt at mounting birds was a Steller's Jay that had been shot along the Willamette River. He showed ingenuity, if not aesthetics, in using a lady's hat pin—pushed into the eye socket and through the skull, the shiny black ball served as an eye on one side while the pin, stuck into a board, supported the specimen on the other. The jay was "presentable enough," but Sutton had neglected to remove the brain, and maggots soon tore the head to pieces.

His oological interests also began to show an advance, perhaps through influence of Vernon Bailey's instructions in the *Handbook*. Sutton collected a Brewer's Blackbird nest with four eggs, and asserting he was through with "amateurish stealing of a bird egg now and then," carefully prepared the set in a professional manner. He began to acquire quite a museum in his room—part of a Steller's Jay specimen (what the maggots didn't want), skins of goldfinches, orioles, crows, nests with eggs, little cards with feathers pasted on. But his father was a restless man, and soon the family was packing again. After just one year in Ashland, and a year in Eugene, Sutton boxed up his museum and moved with his family to Illinois.

George Miksch Sutton left Oregon, but he took more with him than his collection of skins and nests. He had acquired new skills and aspirations. His interest in birds had deepened in Oregon. He learned to prepare study skins and something of mounting specimens. His oological hobby took a more professional direction. And he began drawing birds "in earnest." In Fuertes' plates, first seen in the university's library, he saw what was possible and, almost without distraction, set about to achieve it.

Sutton died in 1982, at the age of eighty-four, but his birds still live. Open any book with his portraits and a bird will look you in the eye and fly from the page, for Sutton gave it life with his brush. And when it sings, you can hear the echo of a Varied Thrush, singing to a ten-year old boy in an Oregon library.[12] —*George A. Jobanek*

Johnson A. Neff (1900–1972)

Western Oregon today, with orchards sprouting in every available flatland, would feel like home to Johnson Neff, who came from and returned to his family orchards in Missouri, having spent time in Colorado as well. Neff came from a family interested in birds: his mother was an active bander, and he published a note about her banding in *Wilson Bulletin* while he was a student.

In between his Missouri and Colorado days, he was a graduate student and teaching fellow in the department of horticulture at Oregon State University, where he produced one of the most detailed and remarkable studies of woodpeckers known at the time (and since). This 1926 thesis was eventually published in 1928 as a monograph by the Free Press of Marionville, Missouri. It bore the long but accurate title *A Study of the Economic Status of the Common Woodpeckers in Relation to Oregon Horticulture*. This provides an example of the kind of research that was becoming more common. It was university-based, used customary scientific techniques, and was conducted by someone intending to be a wildlife professional.

Neff visited friends and his great-aunt's family in southern Oregon during and after his college days. In a 1971 letter to Otis Swisher of Medford, one of the state's active bird-banders, he described one of those visits.

We always had a lot of fun talking about how we found the colony of tricolors north of Klamath Falls. I'd seen them down in the swampy country toward the state line, feeding in fields and in feeding lots, but never northward. I took my family up to the Rogue in, I believe, 1933. Carl and May [Richardson] were then running a small store at Prospect, and Carl and I took off via the "Park" route over to Klamath County.

Carl had mentioned that he had not seen a cowbird as yet, so as we drove down the east side of Agency Lake I was on the lookout for some place with cattle near the road where there just might be a cowbird. Quite a ways down, where the lake curved away from that road as then existing, a herd of such cattle were found and we parked the car and got over into the big grazing area and started to walk down toward the cattle, which would take us well down to the bulrushes and cattails.

In doing so we had to pass under the big powerline, and I was within just a few feet of one of the poles. Just then a bird called out and I did a whirl-about in the air, almost disjointing my back, staring in every direction. Seeing nothing at all, I trotted out a few yards and looked up. There on top of the pole was a tricolored re[d]wing, which was the source of the call I knew so well.

We traced the bird and her colony mates to a spot near the lake in the bulrushes where the beginnings of [n]ests were found. Collected a very few which Carl made up, the only tricolor skins from Oregon in the Biological Survey of National Museum collection at that time.

. . . My great Aunt and her husband moved to Medford area way back before the turn of the century. They had a ranch not too far off the Jacksonville road, up Griffin Creek, and grew apples and cherries, possibly other fruits.[13]

One of the odd lacunae in *Birds of Oregon* (1940) is that Ira Gabrielson, who wrote most of the book, never mentioned Neff's thesis despite having read it and providing some information for it. The woodpecker book remains an excellent source for woodpecker dietary information and was cited in *Birds of Oregon* (2003). Neff became an expert on Band-tailed Pigeons and published a monograph on them for the USFWS in 1947.

THE OREGON FISH AND WILDLIFE COMMISSION

The early twentieth century saw the establishment of Oregon's first state agency focused on wildlife. Game laws had existed for some time—the first were passed in 1872 and, among other things, closed hunting of certain waterfowl from April to July and protected upland game birds during spring months. However, the "Legislature adjourned without appropriating money or assigning anyone responsibility to enforce the laws."[14] In 1887 a three-member Fish Commission was established, which also had authority over upland game birds. The commission lasted one season owing to lack of funding. In 1893 a new entity, the State Game and Fish Protector, was established with the hiring of Hollister McGuire, who drowned in the Umpqua River in 1898 while looking for a new hatchery site.

Early years of the agency were focused on fisheries, but in 1911 the state began an active role in pheasant propagation and bought its first land for wildlife purposes; in 1944 the first full-scale wildlife management area, Summer Lake, was purchased. In his history of the early game commission, Frank B. Wire noted that

> This is a scientific age and the Commission realizing the need for scientific research and investigation has for a number of years paid the Oregon State College for a limited amount of such work. In 1936 advantage was taken of the opportunity to have located at Corvallis one of the biological research units established by the US Bureau of Biological Survey. This work is supported cooperatively by the Biological Survey, Oregon State Agricultural College and the Game Commission. An annual contribution of $6,000 in cash and services is made by the Commission toward this work, which is under the direction of Arthur S. Einarsen, who was sent out by the Biological Survey.[15]

The commission introduced turkeys in 1961. See *As the Condor Soars* for discussion of modern research and management activities by the agency.

CHAPTER 9

Oregon Bird Books, 1901–1939

Alan L. Contreras

Florence Merriam Bailey • Willard A. Eliot • Ralph Hoffmann • William R. Lord • Arthur Roy Woodcock

As more people began studying birds in the lab and observing them in the field, a perceived demand for books about the state's birds led to a series of publications, some of which remain useful today for their detailed information on the status of birds between the late 1800s and the late 1920s. As noted earlier, there had been some local checklists published and a couple of state lists appeared in *Oregon Naturalist*. The varying levels of accuracy found in these lists improved significantly with some of the first state and regional publications.

A First Book Upon the Birds of Oregon and Washington, *by William R. Lord (1901, 1902, 1913)*

William R. Lord came to Portland as a Unitarian pastor. His hardbound book on birds of the Northwest was first published privately by the author. The 1901 first edition[1] was quite small, a 200-page octodecimo (about 4x6 inches) with a hard green cover and a number of black-and-white photographs of mounted birds. It was, however, quite well written in the somewhat florid language of the time, and contained a checklist and "forms" showing what to look for on a bird. In this respect it is a clear precursor to modern books. The information contained in it is generally quite accurate based on what was known at the time, and some of it remains true to this day, for example that the Common Yellowthroat typically arrives in the last week of March.

One unique aspect of this book that was not effectively replicated for Oregon birds until the advent of modern electronically readable sono-grams is that various song types of the Western Meadowlark were set forth on a page in standard musical notation that anyone could play on a piano, at least in terms of pitch and pattern. These songs, augmented with more in the 1902 and 1913 editions, were provided by A. F. Hofer of Salem, son of a German bandmaster who had carefully written down all the Salem-area meadowlark songs he could hear, plus a couple culled from *The Auk*.

The self-published 1901 book was so popular that the Oregon Text Book Commission asked for a reprinting for use in schools. This was done in 1902 in an enlarged 315-page edition published by Heintzemann of Boston. The first edition's meadowlark cover was replaced by a stylized Evening Grosbeak based on one of the illustrations in the book. On the title page Lord's name is followed by the name of the J. K. Gill Company, which operated a department store and bookstore in Portland.

The 1901 first edition claimed to cover Oregon and Washington, but in effect it covered common birds of the greater Portland area. The 1902 and 1913 editions included supplemental material for areas east of the Cascades and the Puget Sound area. They also included suggestions for teaching about birds, size keys, and other useful information. For that era, these were very well-conceived and -produced books. That Lord could produce work of such quality only two years after arriving in Oregon speaks highly of his capability.

Annotated List of the Birds of Oregon, by Arthur Roy Woodcock (1902)

At almost the same time that Lord issued his small book, Oregon Agricultural College (now Oregon State University) student Arthur Roy

Woodcock issued his 117-page master's thesis as a number in the college's agricultural bulletin series.[2] This was a set of papers sent to any Oregonian who requested one. This annotated list was the first attempt to set forth a complete list of the state's birds since Arthur Pope's list discussed in Chapter 5. This laudable goal was carried out in a way that, although perhaps the best that could be done at the time, produced a rather uneven product.

The state-wide list was produced by gathering the existing regional lists that he could find and assembling them, more or less unedited, into a collection. Thus Woodcock included material from active observers such as Bernard Bretherton, A. W. Anthony, and Albert G. Prill, combined with extracts from existing published works such as Charles Emil Bendire's *Life Histories* and Lyman Belding's *Land Birds of the Pacific District*, which was itself a compilation.

The net effect of Woodcock's choice of approach was that Oregon now had, in one place, a solid compilation of what various observers thought were the birds of Oregon. Most of the material was reasonably accurate. Some was less so, and some was either derived from the same sources Belding used or was too generic to serve much purpose. Nonetheless, because of the way Woodcock laid out the material, it was very easy to determine the source of a given "fact" about Oregon's birds.

Unlike the work of Shelton and Neff, Woodcock's compilation could not be described as a scientific work. In a sense, what Woodcock had produced was as much a bibliography of Oregon bird status sources as it was a checklist. Unfortunately, he did not actually list the published sources of much of what he included, perhaps because some of the material was unpublished. Nonetheless, the absence of notes or an actual bibliography is something of an inconvenience for a modern reader. It is necessary to have a copy of Jobanek's bibliography (see Introduction) at hand to figure out where Woodcock probably got his information. Finally, an interesting curiosity is that Woodcock included Oregon's laws affecting birds as an appendix.

A Handbook of Birds of the Western United States, *by Florence Merriam Bailey (editions 1902–1935)*

Although it was not an Oregon-specific book, Bailey's *Handbook* was, in its 1902 edition, the first full-scale "field guide" for the West, something of a companion for Frank Chapman's eastern guide. In its early editions it included verbatim much of A. W. Anthony's list of birds of the Portland area. It was also, along with Bailey's earlier books, among the first to use art by the remarkably talented Louis Agassiz Fuertes. Some of the line art was also exceptionally accurate and detailed. For a more detailed discussion of the history of bird art and artists in Oregon, see *As the Condor Soars*.

The style of the text was a curious mix of technical detail (some, including the measurements, used with permission from Robert Ridgway's 1887 *Manual of North American Birds*) and personal reminiscence, a combination to be repeated in Ralph Hoffmann's more literary *Birds of the Pacific States* in 1927 (see below). Portions were written by Vernon Bailey, Florence's husband and a well-known mammalogist. They spent some time in Oregon, and Florence M. Bailey wrote a couple of articles based on her time in Oregon.[3]

A good biography of F. M. Bailey is *No Woman Tenderfoot: Florence Merriam Bailey, Pioneer Naturalist,* by Harriet Kofalk (College Station: Texas A&M University Press, 1989). This includes her time in Oregon.

Birds of the Pacific Coast, *by Willard Ayres Eliot (1923)*

The early twentieth century was a time during which bird books were in transition. The immense technical manuals for which Elliott Coues, Robert Ridgway, and others had been known in the late 1800s were used by relatively few people interested in birds by the 1920s, as expanding public interest meant that there had to be books that a hobbyist or casual observer could use. Likewise, as ornithology expanded from its base in museums, the detailed descriptions and keys in early references were less connected to what was being studied, and how bird study was conducted.

Among other things, bird books needed illustrations of birds that looked reasonably alive rather than poorly printed sketches of museum specimens on their backs in trays. T. Gilbert Pearson's *Birds of America* had good illustrations of many common species. However, it was enormous, more a desk reference than a field guide, and did not appear until 1936.

Florence Merriam Bailey's books began the trend of more portable guides by using black and white prints of Fuertes art in addition to "museum portraits." The advent of inexpensive color printing allowed for books such as Eliot's 1923 *Birds of the Pacific Coast*, the first bird book written by an Oregonian (Eliot was from Portland) to use extensive color plates as illustrations. These plates were also the first in color to generate wide exposure for an artist then resident in Oregon, R. Bruce Horsfall of Portland. A selection of Horsfall's art appears in the color plates in this book.

With a remarkable fifty-six color plates, this was in some ways the first modern field guide usable for Oregon. In other ways it was a throwback to the Lord-style regional guide, with a somewhat chatty style, though not so anthropomorphic as some early guides. There was a bit more information about behavior (one of the strengths of Bailey's books as well), and the general tone was certainly well informed and clear, if not that of scientific ornithology. This book was nominally about the Pacific Coast, but one reviewer noted with considerable accuracy that the book "fairly breathes of Oregon." It does, and it also showcases a very good artist in Horsfall. His work is often as precise with regard to color and proportion as that of Fuertes, lacking only the exceptional sense of depth and dimension of the senior artist.

Birds of the Pacific States, *by Ralph Hoffmann (1927)*

What a difference a few years made, between Eliot and Hoffmann. Yet the difference is largely of style and technical facility, not concept. Hoffmann was by far the best writer among early field guide authors, with a smooth, impeccable literary flow that can be enjoyed in almost every species account. His style, though informed by earlier literary norms, is very clearly that of the twentieth century, unlike the other writers noted here, and the line art has moved up a notch, too. Illustrator Allan Brooks has a somewhat flat style in his painted bird images, emphasizing habitat, but his line art is wonderfully clear, accurate, and evocative.

This book, too, is not Oregon-specific, and perhaps breathed less of pure Oregon air than did Eliot's, yet some of the descriptions are exactly what Oregon observers experience to this day. We have always been appreciative of his accounts of the shorebirds that breed in eastern Oregon, for example the American Avocet:

> If one's first introduction to an avocet is on its breeding grounds at the muddy border of a slough or marsh, the meeting is likely to be a startling one. Above the shrill clamor of the fluttering terns and the yelping of Black-necked Stilts, the intruder hears a loud *wheep* and becomes aware of a large black-and-white bird flying straight at his head, its *long, slightly upcurved bill* pointed at his face. One has hardly time to note the *cinnamon-brown of head and neck*, as one involuntarily dodges and the bird sails past. (*Emphasis in original*)[4]

This combination of rather high-toned adventure writing and accurate field marks is all but unique to Hoffmann, and the entire book reads this way. It is the only field-oriented North American bird guide of which we are aware that stands on its own as a piece of literature.

CHAPTER 10

Ira Gabrielson, Stanley Jewett, and the Publication of *Birds of Oregon* (1940)

Alan L. Contreras and David B. Marshall

David B. Marshall wrote about his colleagues in a series of articles for Oregon Birds magazine. They are adapted here and remain a first-person story, with some edits and adjustments by the editors.

Serious Oregon bird observers are familiar with the highly regarded *Birds of Oregon* (1940) or its reprint, inaccurately re-named *Birds of the Pacific Northwest* (Gabrielson and Jewett 1970).[1] However, the books' authors, Ira N. Gabrielson (1889–1977) and Stanley G. Jewett (1885–1955), are relatively unknown today. It seems appropriate to describe Gabrielson and Jewett in terms of their personalities and accomplishments, not only because I knew these men well, but also because they contributed so much to ornithology and wildlife conservation. *Birds of Oregon* remains a classic among state bird books.

Ira N. Gabrielson
Formally, Gabrielson was named Ira Noel Gabrielson.[2] To me he was just "Gabe." I enjoyed spending time with him during my employment with the

US Fish and Wildlife Service. One of these instances occurred in the early 1960s when pilot-biologist Ray Glahn and I took Gabe, then president of the Wildlife Management Institute, on an aerial inspection of western Oregon national wildlife refuges. Although Gabe was mum on the purpose of his visit, I strongly suspected it was to view for himself the integrity of the newly established William L. Finley and Baskett Slough National Wildlife Refuges. Both were under attack by the General Accounting Office because they maintained we had no business purchasing uplands with Duck Stamp funds.

Having nominated these areas for refuges and having played a key role in boundary establishment, my credibility was at stake, plus the fact that I badly wanted to see the uplands made a part of the refuges for their terrestrial animal and plant values. Gabe's only response to my prying to see if and what he was going to do about the issue was a gruff, "to hell with them." The issue quieted down. I never found out what role Gabe might have played in preserving the two refuges because he acted behind the scenes, but I learned many years later that John Gottschalk, then director of the Fish and Wildlife Service and a friend of Gabe's, refused to take action on the matter. Julia Butler Hansen, chairperson of the House Appropriations Committee, was quoted as saying, "Well, I guess that's that."

In December 1972, I agreed to transfer to a position in Washington, DC. This provided me the opportunity to really know Gabe and his wife, Clara. On the entry wall was the original of Olaus J. Murie's painting of the Spotted Owl that is the frontispiece in *Birds of Oregon*. Gabe told me that owl came into a camp that he and Murie had on the Oregon coast, and that Murie sketched it on the spot.

Bureau of Biological Survey in Oregon: 1918 to 1935

Gabe yearned to go west after beginning his career in the Midwest and East. In July 1918 he accepted a transfer to participate in prairie dog and other rodent control activities in North and South Dakota and southern Minnesota. This led to his being offered a position in Oregon as rodent control supervisor. "Oregon couldn't be any worse than South Dakota," he said.

In December 1918, he, Clara, and their first child, June, took the train to Corvallis, Oregon, where Gabe was to be stationed near the state extension service at what was then called Oregon Agricultural College (now Oregon State University).

One has to place Gabe's work then in context with the times. A high percentage of the US population was engaged in farming. It was the responsibility of the Bureau of Biological Survey to determine which birds and mammals should be classified as beneficial and which ones should be controlled for agriculture. Gabe was particularly interested in developing and adopting control methods that were not lethal to birds. When Gabe came to Oregon, the Migratory Bird Treaty Act that would form the basis for federal protection of birds had just been signed. The refuge system, while in existence, was not funded or staffed except in a few cases by Audubon Society groups.

Ira Gabrielson in the field. Photo courtesy of the Library of Congress.

It was fortunate that the Bureau of Biological Survey was staffed by research-oriented visionary people like Gabe. They set out to catalog the status and distribution of birds and mammals throughout the country during their off-hours. This began when there were no field guides, optical equipment was rudimentary and very expensive, and the transportation system consisted of railroads, horses, and Model T Fords operating on two-track roads. Gabe wrote of his beginning in Oregon as follows: "Every trip was an adventure, a situation that proved to be true wherever I went in Oregon in those early days. If we didn't get stuck in the mud, or high centered by a rock, the day was relatively uneventful."

In spring 1919 during a trip to Wallowa County, Gabe met Stan Jewett, who was responsible for predatory animal control. Gabe found Jewett to be a capable scientist with shared interests. On this trip they covered much of Wallowa County using a Model T Ford equipped with chains on all four wheels that was driven by a county extension agent named Mac. Of this, Gabe wrote, "Despite the equipment, we were stuck most of the time, and Jewett and I spent most of our time pushing the Ford while Mac drove."

Gabe's first trip to Lake County was also an adventure. The first leg was by train to Bend. Then came a two-day Model T Ford stage trip to Lakeview via Silver and Summer lakes. The Ford hardly got out of first gear because of the rocks and ruts. The return was via Klamath Falls. A horse-driven stage operated between Lakeview and Klamath Falls. This leg took two days with an overnight at Bly.

Eventually Gabe and Jewett came up with the idea of a book on Oregon birds. This required their spending vacations at locales, such as the coast, where their official duties did not take them. Gabe built a family beach house at Devils Lake in Lincoln County. They got fishing boat captains sufficiently interested in birds to take them offshore, provided they furnished the fuel. Gabe's memoirs indicate that *Birds of Oregon* was largely written in the 1930s, and that the first draft was completed in 1936. *Birds of Oregon* was not their first joint writing venture. In 1929 they authored *Birds of the Portland Area, Oregon*, a fifty-four-page monograph.[3]

Gabe became regional supervisor for rodent and predatory animal control for Washington, Oregon, California, Nevada, and Idaho. Three years later, in 1933, regional supervisor positions were established to cover all field work of the bureau, including law enforcement and refuges. Gabe was promoted to this position, the equivalent of today's US Fish and Wildlife Service regional director.

At the time of this appointment, real staffing of refuges began and Civilian Conservation Corps (CCC) workers became available. Among others, Gabe established the CCC camp at what became the headquarters of Malheur Migratory Bird Refuge (now called Malheur National Wildlife Refuge). Up to that time, the headquarters site was almost treeless and without buildings.

In 1934, the famous cartoonist and conservationist Jay N. "Ding" Darling was appointed chief of the Bureau of Biological Survey. The droughts of the early 1930s reduced waterfowl populations to a point that many people thought they would never recover. By July 1934, Darling succeeded in getting emergency money for refuge acquisitions to address the waterfowl crisis.

Gabe recommended first priority be acquisition of the P Ranch, which now constitutes the Blitzen Valley portion of the Malheur National Wildlife Refuge (the original refuge consisted only of Malheur and Harney Lakes). The P Ranch was purchased by early October 1934, at which time Gabe and Jewett immediately departed for Malheur. With the help of Merle Jacobs, the Survey trapper in the area, they opened the gates blocking the Donner und Blitzen River to let water back into the then dry Malheur Lake. Because of the drought, it had little immediate effect, but it was a symbolic action. Gabe wrote, "I don't believe anything that I have done gave me more satisfaction than having a hand in seeing the restoration of this area."

In 1939 Gabe had been offered the presidency of Oregon State College (now Oregon State University). He was tempted, but turned it down feeling he had too much unfinished business within the Fish and Wildlife Service. Gabe worked under Presidents Roosevelt and Truman. He once told me he knew them well, as well as Presidents Eisenhower, Kennedy, Johnson, and Nixon. "Truman was the only one who grew," he once told me. "The others just swelled."

Wildlife Management Institute: 1946–1978

On April 6, 1946, Gabe began duties as president of the American Wildlife Institute, which was soon renamed the Wildlife Management Institute. Gabe received numerous citations, medals, and awards, including several honorary doctorates. He retired from the presidency of the Wildlife Management Institute in 1970, but continued on as chairman of the board until his death at age eighty-seven in September 1978. I feel his greatest asset was an ability to combine natural history knowledge and humor with a refusal to compromise science or principles. It made people listen [a description that could well apply to Marshall, too—*Editors*].

Stanley Jewett

Stanley Jewett was originally from Frederickton, New Brunswick, in Canada. My experiences with Stan began when I was about nine years old in the mid-1930s. My father, Earl A. Marshall, his brother, C. L. Marshall, and other family members were amateur naturalists and early members of the Oregon Audubon Society (now Portland Audubon Society). It was evident to me, even at that age, that Stan's knowledge of birds (and mammals) completely outstripped others associated with the Oregon Audubon Society. In fact, I soon had it figured out that the ornithological knowledge possessed by other society members, other than William L. Finley and possibly Henry M. DuBois, was limited. By this time Gabrielson, the only other really knowledgeable ornithologist in the Portland area, had left Oregon.

Other memories are of Stanley Jewett presiding over the compilation of Portland Christmas Bird Count reports. The participants would meet at a member's home at the end of the count day. Stan would carefully go over all questionable reports, quizzing the participants intensely. He was the final judge. Stan had to be fully satisfied before he would accept anything unusual. He was positive and blunt when it came to a decision, which was often expressed by the words "possible but not probable," and that always seemed to end the discussion.

This was before there were Peterson field guides, at least for the West. Amateur birders did not have much to go on for bird identification, especially if uncommon or rare species were involved. Affordable optics were almost nonexistent, and transportation for some still centered on the city rail transit system. This combination resulted in numerous identification questions, and like beginning birders of today, people often came up with rarities which were suspect.

During Audubon gatherings, Stan would often bring up those conservation issues for which he had very strong opinions. William L. Finley would sometimes be present. As a boy I listened intently to what they had to say. I learned that Stan was highly respected by members of conservation organizations and was "guilty" of getting a lot done through those organizations, a practice that was not always popular with superiors in a government agency. He and Finley worked together on issues, among which was gaining public support for establishment of the Hart Mountain National Antelope Refuge.

Stanley Jewett.

Despite Stan's practice of questioning us on observations, he was not a negative person. To the contrary, he was perhaps the most enthusiastic, energetic, and dedicated biologist I ever knew. He obviously loved his work and all aspects of animal and plant life. Anything new excited him, especially new species for the state.

Stan took notice of me during the Malheur trip (see Chapter 17), and some months later invited my father, my younger brother Albert, and me to his Westmoreland home. The basement of the home was our focus. In it were cabinets containing hundreds if not thousands of study skins of birds and mammals and shelves of scientific journals that included the *Auk*, *Condor*, and *Murrelet*. I subsequently found out that the recently skinned Ross' Geese hanging over the laundry tubs to dry after washing and fat removal did not represent an unusual situation. As my father told me after the first visit, I had become acquainted with a real scientist. This was the first of many visits to Stan's basement.

Albert and I were not the only young people who got into Stan's basement. By the time I was twelve years old, I picked up a birding friend, Tom McAllister, and soon thereafter Bill (William H.) Telfer appeared. Tom went on to be the outdoor writer for the *Oregonian* and Bill became a professor of biology at a Pennsylvania college. Stan took all three, and sometimes four, of us birding. We regularly consulted with him regarding what we had seen on our field trips. What we thought might be a rare bird often turned out otherwise after a consultation with Stan.

It did not take long for Stan to ask, "Where are your notes?" He insisted on field notes and personally looked them over. This was the beginning of my formal education as a wildlife biologist. Stan was intent on making us collectors like most professional biologists of the time. He could skin and stuff a bird in what seemed like seconds with a few flips of his hands and fingers. He was determined to teach us how to make bird and mammal study skins. At Stan's insistence, I had a scientific collecting permit by the time I was sixteen or seventeen. To go with it was a supply of dust shot and a .410 shotgun. Stan personally instructed us on how to make study skins, and especially to properly label them. None of us took to this like Stan had hoped.

Nonetheless, he convinced me that the only way I could get any rare bird sightings accepted and published during that period was to collect them. He was 100 percent correct on this, and up into the 1960s I collected a number of birds for this purpose, most of which are deposited in the US National Museum. Most rare bird records for Oregon during the 1940s and 1950s were published in *The Condor*. The editor was Alden H. Miller of the Museum of Vertebrate Zoology at UC-Berkeley. He routinely refused to publish any sight records that people like myself turned in. Photographic equipment and film speeds at that time were not adequate to document rare birds, and there were seldom other knowledgeable birders or biologists to verify sightings.

I often reflect on the specimen requirement policy for unusual records. It left little to no doubt as to the validity of the identification and provided a permanent record. I can see many of today's sight records or others without published photographic evidence being questioned in the future. Stan used to point out to those who criticized his collecting that he took fewer birds a year than a single Cooper's Hawk.

Stan told Bill, Tom, and me that if I we wanted to enter government service as wildlife biologists or managers, it would take more than a university degree. He stressed summer vacation work, even if it began with trail building. Accordingly, Tom McAllister and I got our first resource agency positions in 1943 at age seventeen during the labor shortage of World War II. We began on work crews in the Fremont National Forest, and then each of us was assigned to our own fire lookout stations. Our summer's bird notes resulted in publication of a paper in the *Auk* titled "Summer Birds of Fremont National Forest, Oregon."[4] Stan was very helpful in seeing this come about.

In 1935, about the time that Gabrielson was transferred to Washington, DC, Stan became the first refuge manager, then called superintendent, for the Malheur Migratory Bird Refuge. However, it must have been on a commuter basis because he continued to live in Portland and held the position for less than a year. John Scharff then took over the refuge, becoming the first resident superintendent.

Stan then became regional refuge biologist for Region 1 of the Survey. The region then included six western states. He held this position until 1941, when he became the first Pacific Flyway biologist. The last two years of government service before retirement in 1949 were as a wildlife research biologist.

One of Stan's last field trips was to Finley Corrals, on the edge of the Gearhart Wilderness in the Fremont National Forest. I heard about it from

"Spike" Armstrong, the district ranger. Spike joined the Jewett party, and once told me that the small quantities of meat removed from small mammals and birds during the skin preparation process went into a stew which was continuously renewed as it cooked over a fire.

Stan received many honors, but perhaps the most treasured, considering the fact he was self-trained, was receipt of an honorary Doctor of Science degree from Oregon State College in 1953. He was elected as a Fellow of the American Ornithologists' Union in 1940. A bronze plaque, erected by the Izaak Walton League in honor of Stan and William L. Finley, stands near the entry to the headquarters of the Malheur National Wildlife Refuge.

As an author, Stan's greatest accomplishments were probably *Birds of Oregon* and *Birds of Washington State.*[5] I can't think of a better way to end this than to quote some words from the editorial that appeared in the *Oregonian* upon Stan's retirement.

> Having enjoyed the acquaintance of some few of the species, we are qualified to define a biologist as a fortunate nature-lover who gets paid for doing something he would much rather do than anything else. Good biologists are born that way.... Retirement doesn't mean that Stan Jewett will stop being a biologist. That's something he can't do. His enthusiasm for a new specimen, or another item of knowledge about the habits of this creature or that, is as boyish as ever.

A NOTE ON THE PUBLICATION OF *BIRDS OF OREGON* (1940)
Alan L. Contreras

For many Oregon observers, Gabrielson and Jewett's Birds of Oregon, published in 1940 by Oregon State College, now Oregon State University, was the anchor point for what they knew about Oregon's birds as they began their own study. The book was one in a series of OSC monographs.

The 1940 *Birds of Oregon* was written entirely by Gabrielson. Jewett provided much of the research and what Gabrielson described as "the endless task of checking records and literature," a phrase all too familiar to those of us who worked on the 2003 *Birds of Oregon*, for which Dave Marshall served as senior editor (see Chapter 18). Their process included handwritten records on individual pieces of paper for each species in each county.[6]

Some of the basic descriptive material was reprinted from the 1921 edition of Florence Merriam Bailey's *Handbook of Birds of the Western United States*. Bailey in turn had developed some of her material from Ridgway's various works. Gabrielson and Jewett also used information from A. C. Bent regarding downy young birds. George M. Sutton's superb series on plumages of young birds had just started to appear in the mid-1930s and covered mostly eastern birds.

Users of the 1940 *Birds of Oregon* are sometimes unaware that the book was in reality somewhat older. The manuscript was completed in June 1935, except for a few stray records. As Gabrielson put it in the Preface, "The mass of available data seemed appalling when the compilation started," and he mentions the help of Adelaide King, who not only typed the complex manuscript but tabulated the card filing system that Gabrielson used for records.

There were several smaller statewide publications issued between Gabrielson and Jewett's book and Marshall et al's *Birds of Oregon* in 2003, the most useful of which were probably the Bertrand and Scott annotated checklist of 1971, the two small Portland Audubon booklets by Harry Nehls and Marshall (which covered perhaps two-thirds of the state's birds) and Gilligan et al's *Birds of Oregon: Status and Distribution* (1994).

CHAPTER 11

The Emergence of the Active Amateur, 1901–1960

Range Bayer, Alan L. Contreras, and George A. Jobanek

Wes Batterson • Overton Dowell Jr. • Reed Ferris • Tom McCamant • Grace McCormac French • Ruth Hopson Keen • Albert G. Prill • Mary Raker • Hilda Reiher • Alex Walker • Ann Ward

We have focused on government surveys, early academic activity, and other study that was variably scientific and occasionally professional in the context of their time and place. Yet in ornithology as in other sciences with a significant field component, determining who was a "professional" and who wasn't was often ambiguous speculation in territorial days and early statehood. It was also not very useful information because this was an era in which college education was rare—degree attainment was low. In 1940, when Ira Gabrielson and Stanley Jewett published *Birds of Oregon*, "more than half of the US population had completed no more than an eighth-grade education. Only 6 percent of males and 4 percent of females had completed 4 years of college."[1] During the nineteenth and early twentieth centuries, anyone with an interest, some kind of reference material, and sufficient dust shot could be an ornithologist.

Although Florence Merriam Bailey, Frank Chapman, and others worked to encourage largely observational ornithology, in practice there was limited equipment, mostly 3x or 4x opera glasses, and no well-illustrated field guides until the 1920s. Professional ornithology was essentially museum ornithology, and even there the people doing the work, for example, Robert Ridgway or Stanley Jewett, had no academic credentials.

By the 1960s large numbers of people who were not scientists and did not want to become professional ornithologists became interested in observing birds. This provided opportunities and concerns for professional ornithologists. On the one hand, the availability of a large number of reasonably competent observers created opportunities for projects that needed a lot of field hours. On the other hand, for some kinds of research there were questions about whether untrained volunteers could provide data that was good enough to be used in meaningful studies. This situation eventually resulted in a national conference co-sponsored by the National Audubon Society and the Cornell Laboratory of Ornithology titled "The Amateur and North American Ornithology" at Cornell University in 1978. This event discussed the best ways to make use of nonprofessionals, in particular in banding and surveys of water birds and raptors.[2]

What happened during this transitional period? Several things. The Audubon Society became a factor in bird study and preservation. As more people became interested in birds, the market for illustrated field guides and the practicalities of color printing converged. For Oregon observers, the 1920s were the great transition decade, as the color-plate field guides of Eliot, Hoffmann, and eventually Peterson surpassed the largely textual guides of Bailey and the monster references of Coues, Ridgway, and Pearson.

What did this mean for bird study? The main consequence was that anyone who could *see* birds could identify them with reasonable accuracy. That meant that local observers with the willingness to keep records and share information became a part of Oregon ornithology. This chapter is a series of mini-biographies of some of the more active and visible people who studied birds in the first half of the twentieth century. Who were these people? Cheesemakers. Farmers. Homemakers. Doctors. Geologists. Just about anyone.

Alexander (Alex) Walker (1890–1975)

Alex Walker was born in a sod hut on the plains of Nebraska on August 11, 1890.[3] After he graduated from high school, his family moved to South Dakota in 1907, when he was seventeen. There he credited Charles Crutchett for encouraging and spending time with him in bird study and giving him his first bird identification book. He learned taxidermy and won prizes for his mounted birds at county fairs. He also photographed birds, their nests, and eggs and collected bird eggs. He published sixteen papers about birds (some with photos), some in ornithological publications. Walker co-authored the *The Birds of Douglas County, South Dakota*, which was printed chiefly for student use in Douglas County schools. It was published in 1912 when he was twenty-two.

Walker and his family arrived in Oregon by April 1912, when he collected a set of Song Sparrow eggs in Mulino, Clackamas County. His photograph of an American Robin nest in Clackamas County on May 4 of that year was published in Arthur C. Bent's twenty-one-volume series, the *Life Histories of North American Birds*.

Walker supplied photographs and egg information for Bent's entire series as indicated in the first volume published in 1919. He and Erich J. Dietrich conducted the 1912 Christmas Bird Count in Mulino, Clackamas Co. He and Donald E. Brown conducted the 1913 Christmas Bird Count there.

In April 1913 Walker collected the first White-throated Sparrow in Oregon since the 1870s. On May 15, along with his father, Ellsworth Walker, and a friend, Max Short, he left The Dalles to go south to Silver Lake and back with a team of horses and a covered wagon on a five-week collection trip.[4] His South Dakota experiences in the field, including photography and knowing how to make bird study skins, were put to use. He collected bird nests, eggs, and birds for study skins; he also took photos, wrote notes, and collected small mammals during this trip. He vividly recalled almost fifty years later how cold it was there every night: each morning the ice in their water bucket had to be broken. Sometime after this trip, the Oregon Fish and Game Department, as it was then known, bought Walker's specimens and notes to add to their collection.

In 1914, William L. Finley, state game warden for the Fish and Wild-life Department, hired Walker as deputy game warden, perhaps because of his trip in 1913. His main duties were to collect and take photos of birds and mammals during his Biological Survey trips in 1914 and 1915. He was sometimes by himself, but more often he was with other much more ex-perienced field naturalists, including Stan Jewett, Vernon Bailey, Alfred C. Shelton, Morton Peck, and Luther Goldman. In 1915, Walker was contin-ually in the field with Jewett from April 1 through June 1. Near the end of the 1915 survey, he went alone to Tillamook County to continue collecting for the survey.

Alex Walker regarded natural history as his chosen profession; cheese making was his livelihood. After the 1915 survey, his time with the Fish and Game Department was over, and he started working with cheese. He did a 1915 Christmas Bird Count in Tillamook and collected and observed birds in Tillamook County from 1915 to May 1918, noting that Snowy Owls were common during the 1916–1917 winter.[5]

After brief army service in World War I, he again worked at cheese making in Tillamook County and married Rosaline Davies in March 1919. During the 1920s, he was employed in cheese making, but he continued to collect birds. In 1920, he did the Christmas Bird Count at Netarts and collected an Emperor Goose on December 31 at Netarts that was the first record for Oregon. In 1922, he collected a Eurasian Wigeon at Netarts that was the fourth record in Oregon. He took a Black-chinned Hummingbird in the Warner Valley in June 1925 that was only the second record for Or-egon. He also authored five papers in the ornithological journal *Condor* and contributed 13 bird photos to Jewett and Gabrielson's 1929 *Birds of the Portland Area, Oregon*.

In 1928, Walker qualified for being a US Bird Reservation Protector, who worked at US Bird Reservations and Refuges such as Lake Malheur and Upper Klamath in Oregon. He then sold his bird collection to the Uni-versity of California at Los Angeles. In 1930, he was still waiting for an ap-pointment to be a Reservation Protector, when he joined the field staff of the Cleveland Museum of Natural History.

Alex Walker at his field desk. Photo courtesy of the Oregon Historical Society.

He collected birds for the Cleveland Museum during 1930–1932 in Tillamook County and elsewhere in Oregon, California, and Arizona. In the summer of 1930, he did extensive collecting in the Warner Valley area and found seventeen new bird subspecies for Oregon. Walker added more than 8,000 bird study skins to the Cleveland collection, with 3,384 of these collected along the Oregon Coast.

The Great Depression put an end to Walker's collecting with the Cleveland Museum. He was employed in cheese making from 1933 to 1951, but birds and natural history were still his main interest. In October 1933, a dead Common Poorwill at Tillamook was brought to him and was the first record for western Oregon. In November 1934, he collected a Snowy Egret at Nestucca Bay that was the first specimen for Oregon. He also helped Reed Ferris band cormorants, Western Gulls, and Common Murres along the Oregon Coast. Walker also participated in Christmas Bird Counts in Tillamook County in 1935, 1936, 1938, and 1939. He additionally contributed to Gabrielson and Jewett's 1940 *Birds of Oregon* that cited his bird records on sixty-nine pages and included thirty-three of his bird or nest photographs. They also cited nineteen of Walker's papers published prior to 1935 (Jobanek found six more for his bibliography) and one paper published in 1935 about Oregon birds.

In 1940, Alex Walker's nineteen-year-old son Kenneth published an article about Bullock's Orioles on the Oregon Coast, which made it a special year.[6] The very next article in the journal was by Alex about a rare subspecies of Horned Lark in Oregon. Kenneth Walker, who attributed his keen interest in natural history to his father, went on to get his PhD in zoology at Oregon State University and taught for six years at the University of Puget Sound and then for twenty-eight years at Western Oregon University.[7]

In 1949, Alex Walker also wrote about the status of Townsend's Solitaire in the Oregon Coast Range. It was based on his collection, sightings by his son Kenneth and Wes Batterson, and by being led to a nest that Peter Walker Jr. (the son of Alex's younger brother Peter Walker) had discovered. In another paper in 1949, Alex reported that Wes had found the first European Starlings at the ocean and commented that the species seemed "to have reached the limits of its westward expansion in the United States."[8]

In 1951, the Tillamook County Pioneer Museum began searching for someone to set up a natural history exhibit.[9] Walker said he "jumped at the chance" because it "was right up my alley." The next year at the age of sixty-two, he permanently abandoned cheese making to start natural history work at the museum. In 1955, he became curator, and Rosaline became custodian. Alex worked on bird and other natural history exhibits and meeting visitors, even after Rosaline passed away in 1961. At the museum, he set up a natural history room that included bird and mammal mounts.

During the 1950s and 1960s, Alex continued to publish articles in ornithological journals, including Wes Batterson's collection of the third Red-legged Kittiwake record in Oregon and the second Northern Mockingbird record for western Oregon. A 1960 article was about Rosaline first seeing and Alex collecting a Rusty Blackbird that was the first verified record for Oregon, and Alex's sighting of a Dickcissel at the residence of Paul Lewis that Mrs. Lewis would not allow to be collected. It was a first for Oregon, but not the first verified record as it was not collected or satisfactorily photographed. At that time, photographic equipment and film were generally inadequate to take photographs of distant and/or moving birds for identification verification, especially in difficult lighting.

By March 1969 Walker had prepared a three-page brochure "Bird Watching in Tillamook County" for museum visitors, which was still available through at least 1982. Its purpose was to educate and encourage interest in birds and answer some basic questions about local birds to museum visitors. It was also given out at OSU's Hatfield Marine Science Visitor Center. In 1975, it was estimated that 60,000 visitors a year had toured Walker's natural history room at the museum.

Alex Walker's last ornithological paper was in 1972 at the age of eighty-two and was about two Least Terns that Wes Batterson had collected in Clatsop County. The Oregon Bird Records Committee accepted this as the first verified Oregon record. In a way, Alex's life had gone full circle. During 1907–1912, he tried to encourage and educate young people and others about birds and natural history with his taxidermy, egg collection, and publication of birds of Douglas County in South Dakota. From 1952 to his death on August 13, 1975, he tried to do the same with his work at the museum. Almost forty-five years after his death, more than nine hundred of Walker and Batterson's bird and mammal mounts are on display in the Alex Walker Natural History Room to visitors of the Tillamook County Pioneer Museum. Their exhibits are also included as part of the teacher's packet as projects to help educate and interest visiting school groups, especially for grade 2–5 students. —*Range Bayer was principal contributor of this segment*

Reed Ferris (1901–1990)

Reed Ferris arrived in Tillamook County in 1917 and met Walker about a year later when they worked in cheese factories and lived in the same boarding house. They later moved apart and worked at different factories in the county. In 1924, Ferris married Margaret Learn and met Stanley Jewett socially. Jewett had kept in contact with Margaret's family after her father, a minister, had married Jewett and his wife. At one point about ten years after they met, Ferris became Alex Walker's supervisor in the cheese factory at Beaver, Tillamook County—they were the only two employees and perhaps spent their spare moments talking birds during the ten years before Ferris moved to Utah.

Ferris stated in later years that Walker had taught him "how to tell the difference between a robin and a seagull." Walker introduced Ferris to Wes Batterson, William L. Finley, and other active field naturalists. Ferris became a good field observer and one of the first people to conduct significant bird-banding in Oregon.

In about May 1930, Ferris worked full-time and started banding birds as a volunteer for the Biological Survey. He continued banding after the birth of his two children and arrival of Margaret's mother to live with them during 1934–1938. Why did Ferris begin banding? Stanley Jewett worked with the Biological Survey and took Ferris on bird field trips. Wes Batterson may have talked about or introduced Ferris to Batterson's father, S. M. Batterson, who lived in Mohler (Tillamook County) and banded waterfowl in 1924–1926. Walker had previously worked for the Biological Survey, had banded a bird in 1913, saw Ferris almost daily and could exchange bird news after they started working together in the same cheese factory in 1933.[10]

During 1930–1943, Ferris banded 8,000 seabirds (mostly Western Gulls and Common Murres) at Haystack Rock (Pacific City), Cape Lookout, and Three Arch Rocks. Ferris published two papers about his Western Gull banding work.[11] Unfortunately, Ferris didn't analyze and publish his Common Murre banding results. Cursory analyses of band recoveries that did not identify Ferris reported that most Oregon murres went north after banding. Some went into British Columbia waters. In 1987, Ferris's detailed recovery results were published. They have been cited since then, including in the currently available *Birds of North America Online* species account for Common Murres.

Ferris also banded 3,042 terrestrial birds. Recovery data of birds banded by him may have been pooled with other banders and analyzed without identifying him. In 1931, Walker taught Ferris how to make bird study skins, and he made 84. One was a Northern Waterthrush caught in one of Ferris's banding traps, the first Oregon record.[12]

Although Ferris left Oregon in 1944 to pursue a career in cheese making, his banding records remain available for study in the USGS Bird Banding Laboratory database. —*Adapted from Range Bayer* (1987)[13]

THE FIRST BANDED BIRDS IN OREGON

Prior to 1920, bird banding was not under government control. Unfortunately, the banding lab's electronic files are only complete back to 1960, so locating the first bird banded in Oregon can't be reliably done. However, the lab does have the banding data from before 1960 for any bird that was subsequently re-encountered. A Brandt's Cormorant banded by Olaus Murie on July 4, 1913, at Tillamook Bay was reported shot in British Columbia, but with an unknown date. It may be the first bird banded in Oregon; it is definitely the first recovery. There are no re-encounters of Oregon birds before 1913 or between the 1913 record and 1924. —*Danny Bystrak, Bird Banding Lab, Patuxent Wildlife Center*

In the book *Discovery: Great Moments in the Lives of Outstanding Naturalists* by John K. Terres and Thomas W. Nason, this very cormorant is most likely memorialized. In the chapter by Olaus J. Murie called "Escape at Three Arch Rocks," quite harrowing in its own right as the writer was briefly shipwrecked on the rocks, is mentioned that he went to Three Arch Rocks on July 1, 1913, and was there a few days. He and L. Alva Lewis, manager of all federal refuges in Oregon, banded young birds on the colony rock. This is a murre colony but also has breeding cormorants. Murie, who had come to Oregon to attend college at Pacific University in Forest Grove, was working for state game warden William L. Finley at the time. —*Editors*

Wes Batterson (1909–2008)

Wes Batterson, born in the Nehalem Valley in Tillamook County, was a well-known game-bird breeder working for what was then the Oregon State Game Commission. He met Alex Walker in the 1930s, but they did not begin working closely together until about the 1950s.[14] Batterson then worked for the Oregon Department of Fish and Game. He knew Stanley Jewett from when Batterson worked at Malheur Refuge, and met William L. Finley after Finley visited Batterson because of the latter's ability to raise grouse in captivity. Batterson noted that his most important friendship and working relationship was with Alex Walker, and they traveled together on trips to Alaska, Warner Valley, Marys Peak, and other areas.[15]

The Tillamook Museum collection includes not only Walker's mounts but those by Batterson, who had been taught by Walker to mount specimens for public display. Batterson went on to mount over 1,000 specimens for the Tillamook Museum, the OSU Hatfield Marine Science Center in Newport, and OSU in Corvallis.

Batterson was asked to help save the declining Hawaiian Nene goose from extinction. Through captive breeding in the 1950s Batterson was able to increase the population from about fifty to more than a thousand. The Nene survives today.

Grace McCormac French (1881-1957)

To some extent, Grace McCormac French, born in the Coos Bay area, is a stand-in, a type specimen if you will, for the many women who kept local notes and became the "bird lady" of their communities. Other early to mid-twentieth-century examples were Olive Barber in Coos Bay, Hilda Reiher in North Bend, Mary Raker of Portland, and Ann Ward in Baker City. The latter three are profiled below.

As was true of all amateurs (and some professionals), the quality of information gathered by these local observers varied. In general they became quite familiar with the local species and the timing of their arrivals and departures. This allowed them to notice birds that were out of the ordinary in their area. For the most part they stayed close to home, and that local focus made their data more valuable than if they had dashed all over the state.

Grace French was a typical farmer's wife of that era, focusing on her local birds in adulthood as time allowed. However, she developed considerable experience and became well known throughout northwestern Oregon for her many public presentations on birds, including radio programs.

A member of the American Ornithologists' Union (AOU) and the Cooper Club, she often welcomed Portland Audubon groups to her farm. Among other visitors were Florence Merriam Bailey and J. C. Braly; she knew most of the major figures in Oregon ornithology of the period, including Jewett, Finley, Raker, Horsfall, and Eliot.

What was her contribution to knowledge of Oregon birds? Primarily her spring migration tables from Yamhill County and some baseline data for the county in general. Like most observers

Grace McCormac French with an owl specimen. Photo courtesy of the Special Collections and Archives Research Center, Oregon State University Libraries.

of the time, limited by poor field guides and worse optical equipment, her local lists contain occasional errors, but they also provide invaluable data about the status of birds in her area.

She donated her personal bird collection to the Museum of Natural History in the OSU Department of Zoology in 1956; her notebooks are also at OSU.[16] — *Range Bayer contributed to this account*

Overton Dowell, Jr. (1879–1963)

Overton Dowell Jr. was born in northwest Clackamas County, Oregon, near present-day Milwaukie, in 1879 and came to the Florence area in 1882.[17] In 1892, he and his father Overton Dowell Sr. homesteaded adjacent land tracts at the northeast end of the Dowell Arm of Mercer Lake in the Bailey Creek area. Until Dowell Jr. sold his homestead, he and others referred to his area as Dowell Ranch.

From March 1908 through September 1912, Dowell Jr. was employed as second assistant lighthouse keeper first at the Umpqua River Lighthouse and then at Heceta Head Lighthouse. He married Mary Hefty in 1909, and they had their first child. They were provided a residence at the lighthouses. After he left the Lighthouse Service, he and his family moved to Dowell Ranch. They had to cross Mercer Lake by boat to come and go from it because there was no road until 1946.

Between 1912 and 1916, he was employed with the Oregon Fish and Game Commission as a game and fish warden; William L. Finley was state game warden at the time. In 1912, Finley appointed Stanley Jewett to build a collection of birds and prepare a manual for preparing and care of specimens. The manual was published in 1914.[18]

Jewett's and Finley's influence may have led Dowell Jr. to start collecting and making bird and mammal study skins while working for the commission.[19] He collected his first bird specimen at Dowell Ranch in May 1914, where his collecting area was limited. He also began sharing his observations with others, including notes to Alfred C. Shelton for his *Land Birds of West-central Oregon*.

After Dowell Jr. left the commission in 1916, he was self-employed as a dairy farmer at the Dowell Ranch. In 1917, he first reported bird sightings for Mercer Lake to the Biological Survey. He collected more bird skins afterward than while he worked for the commission. After he left the commission, he also collected several bird species that were new records for Oregon. In 1919, he made a trip to eastern Oregon and collected what Jewett later examined and determined was an Upland Plover (today's Upland Sandpiper), a first for Oregon.[20] His other first records for Oregon include Black Phoebe and two species that he had originally misidentified, Thick-billed Murre and Cassin's Kingbird.

In 1929, Jewett visited Dowell Jr.'s private collection of "700 or more beautifully prepared specimens."[21] Jewett commented on six bird specimens including a Brown-headed Cowbird at Dowell Ranch in 1925 that was the first record west of the Cascades in Oregon. In 1940, Dowell's unpublished observations, notes, or specimens were acknowledged and often cited in Gabrielson and Jewett's *Birds of Oregon*.

In 1946, Dowell Jr. and one of his younger brothers, Floyd, who owned nearby land sold their properties. The new landowners named these properties Enchanted Valley, and part of this area is now a public hiking area, though hard to access. After the sale, Dowell Jr. retired and moved with Mary to Dowells Peninsula along the south-central shore of Mercer Lake.

In October 1962, Mary, his wife of almost fifty-three years, passed away. Mary had helped preserve and maintain the collection. Three months later he donated his collection of 800 bird study skins and field notes to Siuslaw High School in Florence. Seven months later he passed away. In 1970, the high school sent his bird specimens and notes to the Museum of Natural History in the Zoology Department at Oregon State University.[22] This museum later transferred the bird specimens to OSU's Department of Fisheries and Wildlife and his field notes to the Western Foundation for Vertebrate Zoology. Some of the specimens were later lost. —*Range Bayer contributed most of this segment.*

THE VALUE OF RETAINING SPECIMENS

On August 4, 1935, Overton Dowell Jr. collected a kingbird at Mercer Lake on the central Oregon coast, where he lived. Dowell had seen Western Kingbirds there before, according to Alfred C. Shelton's 1917 publication, and Dowell's own notes show that he was confused about what he had found, as it was first called a Western Kingbird, then eventually a Cassin's Kingbird.

The specimen was eventually deposited in the museum at Oregon State University. When the Oregon Bird Records Committee wanted to examine the specimen in the 1980s (it was the only Oregon record of Cassin's at the time), it could not be found. It was apparently destroyed when many OSU zoology collections were stored improperly. What could be found were three published testimonials as to what the bird was.

First, Stanley Jewett had published a brief note in *Condor* in 1942 stating that he had seen the specimen and implying that the bird was correctly identified. Dowell himself had contacted Alden Miller at the University of California and apparently sent him the specimen for comment. Miller responded:

> There is not the slightest doubt regarding the identity of the Cassin Kingbird. It is a bird in typical juvenal plumage and agrees exactly with a Cassin Kingbird in similar plumage in our collection.

This suggests that the specimen was sent to Berkeley, but by the early 1970s it was back at OSU, where Ralph Browning saw it while working on an article. Thus although the specimen itself has unexpectedly flown, we have three expert opinions that it was a Cassin's Kingbird. Upon that basis the record was included in Marshall et al., *Birds of Oregon* in 2003, though it is not accepted by the Oregon Bird Records Committee (OBRC).[23] Oregon has since had one more record of the species, quite alive, in November 2001.

Dowell's own notes, copies of which were provided to the editors by Range Bayer, refer to the bird as having a deeply forked tail, and Dowell or someone else eventually scratched out the word Cassin's in a specimen list and wrote in "Tropical." The August date would be very unlikely for that species, which occurs in coastal Oregon in late fall. This case illustrates the value of retaining specimens in good condition.

Tom McCamant (1901–1986)

Tom McCamant, a Congregational minister, grew up in the Portland area. He started birding by age thirteen, spent part of his career in Medford and retired in Salem. He was instrumental in establishing the first local chapters of the National Audubon Society and Christmas Bird Counts in Medford and Salem. Few people were reporting birds from the Rogue Valley when McCamant went there in the fall of 1953. He writes that he

> was immediately delighted with the birds I saw on the four acre tract where our home and the church were located. Gradually I expanded my birding interests to the whole of Jackson County and found good birding companions in Maj. Gen. Joseph Hicks and Ralph Browning,

the latter not yet in high school when I first met him, but an enthusiastic birder.[24]

Among the notable records that McCamant found were some of the earliest reports in southwestern Oregon of Tricolored Blackbird, Black Phoebe, Black-billed Magpie, and Blue-gray Gnatcatcher. Browning went on to become one of the nation's premier avian taxonomists and, upon retiring back home to the Rogue Valley, mentioned McCamant's mentoring and friendship in his memoir *Rogue Birder* (2019).

Salem observer Dona Horine Bolt first met McCamant when she was twelve, and notes that "he was such a wonderful mentor. There aren't many parents who'd let their youngster go off in a VW bug every Saturday to go see birds, but mine did."

Mary Raker (1903–1997)

Not all young bird observers are boys. When William S. Raker took his daughter Mary to a meeting of the young Audubon Society of Portland in 1913, he probably didn't realize that the spark of bird study could affect her. It did, and she went on to become a well-known figure in Portland bird circles, as well as the first teenage girl for which we have much information who became a "bird kid." An article in the *Oregonian* noted that she went on to drag her father further into outdoor activity:

> As a concrete example of the interest that is being taken in Portland today there is the story of Mary Raker, 17 years of age, a graduate of Franklin high school last spring and now taking her first year at Pacific university. Children nearly all take a great interest in birds and animals and little Miss Raker pleased her parents greatly by early showing somewhat out of the ordinary talents in this direction. Until she was 12 years of age she spent a great portion of her time just studying birds as she could by observing the most common ones, sparrows and saucy robins, on the Raker lawn. Then one night came the memorable visit to one of the Audubon meetings. Miss Raker, 12 years old at this time, sat entranced.
>   [W. A.] Eliot heard of Miss Raker and coached her, managing to

inculcate some of his love in the already fertile mind of the little lady, with the result that in the past five years Oregon has seen the development of a child marvel in bird lore, a girl who is already attracting national attention from her aptitude and learning, attained in the pure bird love of being able to attract the feathered mites with the result that they come to her and will even light on her hair and on her hands. Her understanding and sympathy coupled with her exceptional observation powers and self-confidence have made her one of the foremost authorities in the northwest and she is now being sought after for platform talks which have uniformly proven highly interesting.[25]

In 1919 and 1920 she was on Oregon's only Christmas Bird Count (CBC), at Portland. One year she had her own team with two female companions. At age sixteen she had identified 154 species and had completed "the university course in ornithology," perhaps the course offered at that time by correspondence with John Bovard from the University of Oregon.

No less a figure than Ira Gabrielson commented on Raker's identifications, as in this short note from *Bird-Lore* in 1922:

Mrs. W. P. Jones reported a strange bird to members of the Audubon Society and later Miss Mary Raker visited the place and identified the bird as the Chinese Starling (*Acridotheres cristatellus*) [*Now known as the Crested Myna.—Eds.*], a bird with which she was familiar from observations made at Vancouver, B.C., where there is quite a colony of these birds. The writer has visited the locality twice (on February 5 and 6) and carefully watched this bird, and agrees with Miss Raker in her identification. Probably this bird is either an escaped cage-bird or a wanderer from the British Columbia colony. It is quite shy, although it frequently visits the feeding-station. . . . Realizing the necessity for care in basing first records on sight identifications, the writer hesitates to record this bird formally. However, there does not seem to be any chance for mistaken identity of this curiously crested bird. While this is its first known appearance in Portland, there is no reason why it should not eventually spread over the Northwest from the established colony in British Columbia. —*Ira N. Gabrielson, Portland, Oregon.*[26]

Mary Raker (second from right), 1920–1921. Raker spent at least two years at Pacific University. Photo courtesy of Pacific University.

According to Tom McAllister's history of the early years of Portland Audubon, "Mary Raker was a birding prodigy who joined the Society in 1913, and at age fourteen was giving illustrated lectures. She had her own color slides for a program presented to agricultural groups on 'The Commercial Value of Birds.' *Bird-Lore* magazine called her 'a child marvel.' Years later on returning to Portland she was an [Audubon] board member and married Frank Bartlett, the 1952 president."[27]

Albert G. Prill (1869–1958)[28]

The Yellow Rail (*Coturnicops noveboracensis*) was first reported in Oregon by Dr. Albert G. Prill, who collected a female near Scio, Linn County, on February 1, 1900 (see Chapter 14). This specimen, discussed in *Birds of Oregon* (1940) was mounted and donated to the University of Oregon Museum of Natural and Cultural History in 1919, where it remains. He also contributed a mostly reputable list of birds from the Scio area to Woodcock's 1902 statewide list.

If that were all we knew of Prill, we'd say thanks and move on. However, after a period of quiet until about 1920, he started reporting birds on a regular basis. Some of these reports, for example, Morcom's Hummingbird (a subtropical species now known as Bumblebee Hummingbird) and Cassin's Kingbirds where Western was the regular species, caused his reputation to sink. These records were unfortunate in that they burdened Prill with a reputation as untrustworthy when for the most part his observations were credible. They detract from his extensive coverage of Linn County, and his pioneering efforts in Lake County.

In April 1925, Joseph Grinnell, editor of the *Condor*, wrote to Ira Gabrielson about Prill:

> Will you please tell me confidentially what you think of Dr. A. G. Prill—
> that is, as to his reliability. I have had contributions from him for *The*
> *Condor*, have sent them back on the basis of their illiteracy. . . . Would
> any contribution of fact from him be authentic?

Gabrielson responded on April 15. "In reply to your letter of April 6, regarding Dr. A. G. Prill, let me say that I do not know the man personally, but only by reputation and by his contributions to the *Oologist* and *Wilson's Bulletin* [*sic*] which I have seen. On the basis of this I would not put any great credence on any of his contributions, although on a question of some common bird, he probably would be entirely reliable. Anything out of the ordinary in the way of identifications I would question very much." Grinnell rejected Prill's manuscript. —*George A. Jobanek contributed to this account*

Hilda Nieme Reiher (1908–1976) and Olive Barber (1889–1972)

Hilda Reiher (pronounced "rear") was born in the Coos Bay area and as a child had a keen interest in birds.[29] She was almost sixteen years old in January 1924 when she started recording the arrival dates for hummingbirds, swallows, and thrushes. In 1930, she took an ornithology class from the University of Oregon[30] and started recording arrivals and departures for most other bird species. Willard Ayres Eliot (then managing Audubon

House in Portland) was her most valuable teacher about birds. In 1934, she married John Reiher. She was an elementary school teacher for twelve years and taught about birds in her classes.

Her colleague, Coos County teacher Olive Barber, authored an elementary eight-page *Birds of Coos County* for school use in the late 1930s.[31] At the start of the school year, it suggested a bird notebook as a class project. It could include a bird calendar, daily notes, observations, drawings, and writings with a grade or reward at the end of the year. Barber wrote: "Mrs. Reiher, teaching at Kentuck made quite an event of this, having outside judges which lent it an atmosphere of importance dearly loved by the children." Reiher also sponsored nature study groups for teachers (*The World*, Coos Bay, 10/20/1938).

The manuscript of Gabrielson and Jewett's *Birds of Oregon* was completed in June 1935. It did not mention Reiher, but it does not appear that she was intentionally omitted. She had not published any papers about birds before 1935 and there is no indication that she shared her sightings with them or was an observer for the Biological Survey. Olive Barber lived in the Eastside region of Coos Bay, at that time at least as remote as Reiher in the Glasgow hills, and like Reiher didn't collect any birds, but Barber's 1934 publication about the behavior of Screech Owls was cited. Barber would also have been known to Gabrielson and Jewett, since she reported her sightings to the Biological Survey starting in 1930, and they included such reports.

In 1939, Reiher first sent some of her Coos Bay bird sightings (including a flock of at least 200 Western Bluebirds) to the Portland Audubon Society's newsletter, the *Audubon Warbler*. At about the same time she joined the Portland Audubon Society and remained a member to her death. She apparently seldom sent them her bird notes. Bayer only found them mentioned in eleven newsletter issues, with only one or two per decade during the 1940s and 1960s and only four in the 1950s.

In the early 1940s, she was the Audubon organizer for Coos County. She led and organized field trips for teachers and also led bird field trips for the general public. She was also leader of a bird club in the local 4-H (which was a program for youths).

From 1947 to 1970, she and John owned and operated a myrtlewood business in North Bend. She took her 4-H club at least once on an eight-hour Christmas Bird Count of the North Bend area in 1949 that recorded sixty-six species. She evidently didn't report her counts to National Audubon, whose first Christmas Bird Count in Coos County (Coos Bay, which she attended) started in 1973. She may have kept her count tallies in her field notes.

After the myrtlewood business was sold in 1970, she retired at the age of sixty-two. She then more often gave presentations about birds at meetings and luncheons and led or organized more field trips (including three-day bus trips to Malheur Wildlife Refuge). Reiher taught at least six free classes about birds for four to six weeks that were open to the public. She contributed field notes more frequently to the *Audubon Warbler* and *North American Birds* and had more letters or notes about birds in the Coos Bay paper. However, she still didn't publish any ornithological papers. Contreras published some of her significant field notes that had been found in 1982.[32]

Reiher had started recording arrival dates for some species in 1924. She didn't record arrivals every year for every species that she saw or others reported to her, but she had long-term arrival dates for many species, including forty-seven years of arrivals for Rufous Hummingbirds, thirty-five years for Swainson's Thrushes, and twenty-two years for Barn Swallows.[33]

In his *Birds of Coos County* (1998) and in an article about Anna's Hummingbirds, Contreras published part of the limited amount of Reiher's notes that Bayer had copied or transcribed. The notes that remain suggest that Reiher was a thorough and careful observer. Contreras met her in the early 1970s, at which time she was elderly but still active, and his personal impression matched what her surviving notes indicate. Examples of her reports salvaged by Bayer include the following:

Am. White Pelican: Records for Coos Bay area on 18 Jan. 1932 and 6 on 1 Feb. 1934.

Brown Pelican: One near Coos Bay on 20 Jan. 1933.

Great Egret: At Coos Bay area, Reiher recorded them in late Dec. of 1935 and from Jan. 3–8 in 1934–1936, 1938, and 1940. She also found them in late Dec. of 1941 and 1943 and early Jan. of 1945 and 1946. 2 at Coos Bay on 12 Nov. 1961 (1962 *Audubon Field Notes* 16:66). Gabrielson and Jewett note only one record for western Oregon. However, their data cutoff for *Birds of Oregon* was 1935, by which time the book was largely written, so these reports may simply have come too late for inclusion, or have not been communicated outside the south coast region. Reiher was still keeping notes late in life, note "72 were at Coos Bay area on 15 Aug. 1975."

Tundra Swan: 200 were at Coos Bay area (Pony Slough) on 18 Jan. 1932 and 300 on 20 Feb. 1932. There were "no coastal records south of Clatsop Co." according to G & J, which seems a significant oversight of local data except that inter-regional communication was quite limited until *Audubon Field Notes* began expanding in the 1950s.

Snowy Owl: One in Coos Co. on 2 Feb. 1932.

Anna's Hummingbird: G&J note that there were no records for Oregon. Reiher's record for the Coos Bay area on 17 Dec. 1944 was the state's first, though not published until 1999.[34] She also observed them on 27 Nov. 1964, and in 1971–1973.

Painted Bunting: At Coos Bay area (Glasgow) on 10 May 1948 and 9 May 1958.

—Range Bayer contributed most of this account

Dave Brown (1934–2020)

Dave Brown was the dean of living Oregon bird observers until his death on August 2, 2020, as this book was being prepared. He was active in birding starting in 1944 as a boy in the Eugene area. His bird study began as did so many people's, with a teacher, Mrs. Dolan, who handed out bird cards that had to be colored. As a teenager he started attending meetings of the recently formed Eugene Natural History Society.

Brown's viewpoint of the status of some Lane County birds is unique because his father was the Corps of Engineers engineer who designed the Fern Ridge Reservoir complex in west-central Lane County. Therefore, young Dave not only had the opportunity to find a Common Poorwill in the south Eugene hills, but to see the Long Tom marshes as they were when Shelton was active. The reservoir began filling in 1941, and Brown was able to see the rare Wood Sandpiper there in 2008, bringing his observations full circle. His childhood explorations yielded a Canyon Wren on top of Spencer Butte with two Rock Wrens; the latter species has subsequently bred there at least twice.

There was a bird club in Eugene when Brown was a teenager, and he was able to spend field time with pioneer bird observer Ben Pruitt of Thurston as well as work in the University of Oregon museum helping Arnold Shotwell with taxonomic sorting, much as George M. Sutton had done for A. R. Sweetser. His youthful connections included meeting Alex Walker. Brown participated in the first placement of Purple Martin boxes in North Bend and the Florence area. In later life, Dave Brown's knowledge of the back roads of Lane County allowed him to contribute to *Birds of Lane County* (Oregon State University Press, 2006) regarding the best ways to look for birds in the eastern Coast Range.

Ann Ward (1917–2006)

Ann Ward came to Baker County, Oregon, in the 1950s with her doctor husband. Her contributions to Oregon ornithology were many, with such highlights as helping define the range limits of the Bushtit as it moved slowly into northeastern Oregon. Contreras interviewed her at her home in Baker City in 1994, discovering among other things that she had kept her bird notes in her private diary along with material that she did not want the public to see, so her field records were lost. However, she was a regular contributor to *Audubon Field Notes* and its successor *American Birds*.

She recalled corresponding with a skeptical western Oregon field notes editor when a Northern Parula arrived in her yard in early October 1972 and stayed for the fall. She described the early years of Baker County bird study as requiring a few innovations, such as her special list of "the birds that could be here." This was used mainly to screen oddities reported by neophyte observers from the Baker County and Baker Valley Christmas Bird Count (CBC) records, notable for their lack of obvious errors over the three and a half decades that Ward was their screener. Ward started the Baker Christmas Bird Count on December 30, 1956; it was the first count in far eastern Oregon and is the longest-running count east of the Cascades except for Klamath Falls.

Ruth Hopson Keen (1906–1998)

Among mid-century bird observers, Ruth Hopson Keen stands out for several reasons. She was the first woman to serve as a ranger naturalist at Crater Lake National Park (1950–1951), moving into that role after serving as an instructor at the Crater Lake Field School of Nature Appreciation, a five-week course offered by the University of Oregon for the first time during the summer of 1947. In the field school role she worked at Crater Lake with Donald Farner, later president of the American Ornithologists' Union (AOU) and editor of *The Auk*, who included her first park record of Forster's Tern in his *Birds of Crater Lake National Park*.

She was also one of few women with a science doctorate in 1946, fewer with a PhD in geology and even fewer with that background who worked in the true backcountry of the high Cascades. Keen is perhaps best known for her extraordinary twenty-six-year photographic study of the Collier Glacier on the South Sister, begun when she was a high school teacher in Eugene in 1936.

Finally, as is often true in the field sciences, her contributions frequently fell outside the field of her degree: she was a good field ornithologist and a good botanist as well.

Ruth Hopson was born June 19, 1906, in Oklahoma and moved to Oregon before she was five. A graduate of Marshfield High School, she graduated from Oregon Normal School (now Western Oregon University) in 1926. She went on to earn a BA and MA from the University of Oregon and her PhD from Cornell University in 1946. She taught in Oregon public schools from 1927 to 1939 and taught for the Division of Continuing Education, Oregon State System of Higher Education from 1941 to 1971, mainly geology, natural history, and conservation.

A woman of exceptional stamina and field skill, her command of botany[35] and geology are better known than her bird study, but her bird field notes from the 1940s, found in an antique store almost twenty years after her death, provide one of few windows into the status of birds in rural Oregon in that poorly lit period, when many male observers were in the military. These typed and handwritten records are in most cases sufficiently accurate that they provide useful data seventy-five years later about distribution and timing of movements, for example two precisely located records of Green Heron at a time when it was not well known in Oregon. Many of these records have been entered into eBird through the diligence of Paul Adamus. The originals have been deposited with the rest of the large Keen archive at the University of Oregon library, where they await a well-deserved full biography.[36]

CHAPTER 12

Teaching about Birds

Alan L. Contreras, George A. Jobanek, and David Vick

Jim Anderson • Gordon Gullion • Carl Richardson • Charles Quaintance • Classroom Activity • Expanding Refuge Work • Audubon Society Expansion

Material on academic ornithology and bird conservation work at the university and agency level is primarily located in *As the Condor Soars*. However, a brief introduction to the field and biographies of some exemplary early ornithologists who were engaged primarily in educational work is included here.

By the mid-twentieth century, bird study had entered an expansion phase, primarily owing to the advent of professional study in wildlife management and, by the 1960s, a greater governmental interest in bird protection. Faculty at many universities were studying birds. Oregon State University was building its reputation in the field, but smaller Oregon colleges also had notable ornithology work being done.

In addition to those whose modern work is discussed in *As the Condor Soars*, Elver Voth taught for decades at George Fox University (succeeded in the 1990s by Don Powers), and Lowell Spring taught at Western Oregon University. Herb Wisner (still observing at age ninety-nine as this book was in preparation)[1] and Dan Gleason taught field ornithology for non-majors at the University of Oregon starting in the 1960s. Willamette University, home of Morton Peck in the early early twentieth century, continues its long history of bird study with David Craig, whose work on Caspian Terns and other birds is discussed in *As the Condor Soars*. Southern Oregon University also has a long tradition of ornithology work, from Carl Richardson (see below) through Frank Sturges, Steve Cross, and Stewart Janes, who retired in 2019.

Some community colleges have routinely offered a bird course, for example, those by Ben Fawver at Southwestern Oregon from the 1960s through the 1980s and Floyd Weitzel and Joe Russin at Lane from the 1970s through the 2010s. Jim Moodie at Central Oregon Community College notes that he teaches a version of biology 103, ecology, in which he takes students on field trips to learn the local birds. Bret Michalski teaches the free-standing Survey of Northwest Birds (Fish and Wildlife 212) there in the spring.

Bird Study in K–12 Education

We often forget about the contributions made by K-12 teachers to our knowledge of birds, as well as their role in encouraging young observers who may become professionals or simply knowledgeable citizens. In Oregon we have had many examples of teachers whose work has been valuable and inspirational. South Eugene's Jack O'Donnell studied warbler foraging patterns in the 1970s; Bill Hunter from the Eugene school district worked in college studying Black Swifts; his experiences are noted in Rich Levad's *The Coolest Bird* (2007), a book on early attempts to find the nests of this elusive species. A specimen donated by Hunter to the University of Michigan was one of the first ever found.

Some high schools help with bird surveys in their area. Steve Brownfield from Heppner operated the Ruggs-Hardman Christmas Bird Count (CBC) with his high school biology class in the 1960s through the1980s. Among his students in the 1970s was Greg Green, who became a professional biologist and wrote the Burrowing Owl account in Marshall et al., *Birds of Oregon* in 2003.[2] In Enterprise and Halfway in far eastern Oregon, students learn hawk identification and have assisted for many years with Raptor Route surveys (see Chapter 13) .

Diane Cavaness had a bird club at her middle school in Brookings in the 2000s, as does Scott Deckelmann at Scappoose High School today. Deckelmann's student Luke Suchoski, president of the school bird club, notes that

My school has a month-long period where non-traditional classes are taught. Birding is one of the classes that are offered. Birding class is taught by Mr. Deckelmann, who has an enthusiastic and contagious passion for birding. In the class, we are taught how to use eBird, how to make useful observations, how to make a field sketch, and bird anatomy. Most of the class consists of going out and birding on the school campus, and we take five or six field trips throughout the course. The students are engaged and focused on learning how to become a better observer.

The class introduces teens to birding and helps them spend more time in nature. The course helps nurture observation skills, which can help students see their environment around them and how humans impact it. I think this is important because the more aware we become about our environment, the more we can change daily lifestyles to lessen negative human impact on the environment. If it weren't for the bird class offered at our high school, I would not know anything about birding. The class got me and many of my classmates hooked on birding and the class helps teach people what birding is.

The moment when you know that you've made the transition from a casual birdwatcher to a birder is when you can identify a bird by its behaviors. I first experienced this, in my second term of birding class, when a small bird flitted around some pine trees. I immediately knew it was a Dark-eyed Junco and upon closer inspection I was right. In the past year, I have seen my friends get hooked on birding while taking birding class. I can see that they shared the same amount of excitement that I did when you find out how fun birding is.[3]

Chuck Gates taught advanced biology at Crook County High School in Prineville from the mid-1990s through 2013. Classes included learning a small number of local birds in class each week. Students later took field trips and kept records of what they saw. A small school bird club grew from these classes. Larry Thornburgh included birds in his biology classes at Marshfield High School in Coos Bay in the 1970s through the 1980s as did Russ Namitz later. Catlin Gabel's Paul Dickinson took students to Malheur National Wildlife Refuge and on local trips for decades, as have teachers at Portland's Reynolds High School. Mike Patterson in Astoria, James Billstine at Garibaldi, and others have involved students in bird study. Of Dickinson's class, Oregon observer David Bailey notes:

> All semester we would meet at various places around the Metro region an hour and half before the start of school. We always went to Malheur and stayed in a dorm at Malheur Field Station. My experiences there were fundamental to the person I have become, birder and biologist and otherwise.[4]

Those who start out as "kid birders" sometimes become university-based ornithologists or are employed in other academic ornithological work. Oregon has produced many such: for example, Gordon Gullion (Eugene High School; MS, University of California, Berkeley; faculty at Minnesota), Ralph Browning (Medford High School; Southern Oregon University; US Museum), Chris Butler (Beaverton High School; PhD, Oxford; faculty at Central Oklahoma), Teresa Wicks (Hidden Valley High School, Josephine County; PhD, Oregon State University; biologist at Malheur National Wildlife Refuge for Portland Audubon Society), David Swanson (Silverton High School; PhD, Oregon State University; faculty at South Dakota) and David Craig (Scappoose High School; PhD, University of Colorado; faculty at Willamette University). In addition, many work for fish and wildlife agencies.

We have profiled a small number of college-level teachers below, by way of showing the kind of work being done in Oregon during the mid-twentieth century.

Gordon Gullion (1923–1991)

By the early twentieth century, Oregon was growing its own ornithologists from youth on a regular basis. One of the first to become a professional was

Gordon Gullion, born in Eugene. When about twelve years old, he became interested in birds, and spent much time observing the birds of Eugene, as well as coastal locations such as Florence, Yachats, and Coos Bay. In a letter to George Jobanek in 1975, Gullion confided some of his favorite birdwatching locations. "My dad owned a tract of forest and farmland south of the McKenzie River south of Walterville, and I spent many, many Sundays there, in the Coburgs, the Spencers Butte area, and on many hiking trips with the Obsidians [a Eugene hiking club]."

Gullion received his BS degree from the University of Oregon and MS from the University of California, Berkeley. Both before and after his university career, he systematically investigated, with companions, the distribution of birds in the southern Willamette Valley. As he wrote, "In the years from 1946 to 1948, alone, or with Arnold Shotwell, Ben Pruitt or Fred Evenden, I really scoured that country, nearly every weekend, from West Point Hill to Cottage Grove, west to beyond the Fern Ridge Reservoir."

In 1947 Gullion began publishing notes on his discoveries. These included the first breeding record of the Brown-headed Cowbird west of the Cascades and significant distribution records of Wrentits and Yellow-headed Blackbirds. His research culminated in 1951 with the publication in *The Condor* of his thorough and still useful article, "Birds of the Southern Willamette Valley, Oregon."[5] That paper included records of many local observers. For most species he gave a percentage value indicating relative abundance, determined by comparing the numbers of days a species was seen to the number of days a species "might have been recorded."

Gullion listed 210 species, plus 2 species he considered misidentified on Eugene Christmas Bird Counts. He included historic records of Yellow-billed Cuckoo, Spotted Owl, and Great Gray Owl, and records of introduced species such as Gray Partridge, Golden Pheasant, Guinea Fowl, and Wild Turkey. He listed the American White Pelican, Double-crested Cormorant, Great Egret, and Black Tern as common summer birds at the then recently-constructed Fern Ridge Reservoir.

Gordon Gullion went on to bigger accomplishments after he left Oregon. He joined the University of Minnesota's Cloquet Forestry Center in 1958 and became an expert on the Ruffed Grouse. —*George Jobanek*

Carl Richardson (1886–1970)
Southern Oregon had its share of early twentieth-century bird study. One of the most active observers was bander and specimen manager Carl Richardson, who taught at what is now Southern Oregon University. He was one of the first bird-banders in southern Oregon, active beginning in 1926 and ending only with his death in 1970. His later work overlapped with that of Otis Swisher, another long-time Rogue Valley bander.

Richardson also knew Johnson A. Neff, discussed in Chapter 8, and did some field work with him. His presence in southern Oregon overlapped at the end with that of Otis Swisher, Frank Sturges, and the youthful Ralph Browning. —*Alan L. Contreras*

Charles W. Quaintance (1906–2004)
Charles Quaintance, originally from Philadelphia, taught biology at what is now Eastern Oregon University, then Eastern Oregon College of Education, from 1940 until about 1959. With degrees from Arizona, California-Berkeley (MS in ornithology, 1934) and a doctorate from Cornell, he brought exceptional background in bird study to northeastern Oregon.

During his time in Union County he was an active observer and published several records, including a record of Starling from 1946, early nesting records of Starlings, breeding records of American

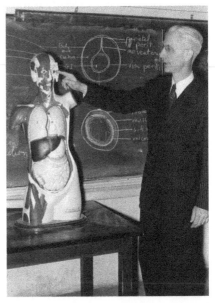

Charles Quaintance. Photo courtesy of the Eastern Oregon University Library Historic Photograph Collection.

The Eastern Oregon University Nature Club, 1949–1950, from the 1950 Mountaineer yearbook. Professor Charles Quaintance at right rear. Photo courtesy of the Eastern Oregon University Library Historic Photograph Collection.

Redstart in Riverside Park in La Grande and a Yellow-billed Cuckoo specimen found in his yard, all published in *Condor*.

His reports were among very few from mid-twentieth-century observers in rural Oregon, as most reports outside the cities were from staff at Malheur and the few other refuges that existed at the time. Thus his 1948 record of a cuckoo was only the fourth for Oregon and the first since 1910, except for a three-year period 1923–1925 when they were apparently common in bottomlands near Portland.[6] Likewise his reports were the first of European Starlings breeding in Oregon.

In addition to his bird records, Quaintance was active in work to preserve the Minam wilderness. One of Oregon's more colorful bird enthusiasts, he spoke Spanish, Arabic, Russian, and French, and moved to France at age seventy-eight to research Holocaust survivors. —*Alan L. Contreras*

Explaining Birds to the Public: The Glorious Flight of Jim Anderson
Jim Anderson (b. 1928) of Deschutes County, Oregon, has been one of the
state's iconic "public biologists" since the 1950s. He grew up in an extended
family in Connecticut that featured uncles who took him out to see what was
happening in obscure corners of the landscape. Among other experiences,
this included acquiring a pet crow whose name was Joe until she laid eggs
and became Josephine. As with so many of his generation, his "field guide"
in the beginning was T. Gilbert Pearson's five-pound *Birds of America*.

Anderson, who arrived in Oregon by motorcycle in 1951, is known
in part for his decades of banding work, focusing on eagles and including
Great Gray Owl. He continued to help in banding raptors until age ninety-one. Anderson worked as the first staff naturalist at the Oregon Museum
of Science and Industry (OMSI) from 1960 to 1968 and also helped to
found the Children's Zoo in Portland. He then became staff naturalist at
Sunriver, a planned resort community south of Bend, where he remained
until 1977, developing the resort's conservation and outdoor education
plan. Jim then went to Arizona to work at Ramsey Canyon until 1980.

*Jim Anderson showing a Great Horned Owl to Oregon Zoo president Hillman
Lueddenmann Jr. at OMSI in 1966. Photo courtesy of the* Oregonian.

Jim's skills as a writer led him to write on natural history subjects for the *Oregonian*, the *Bend Bulletin*, the *Sisters Nugget* and, since 2001, a regular column called "The Natural World" for the *Bend Source*. His book *Tales from a Northwest Naturalist* (1992) includes many of his best stories from a lifetime in outdoor education, including training his domesticated Great Horned Owls to harass visiting insurance salesmen. One of the owls he raised harvested the neighbor's wandering cat; Jim told the neighbor that the cat had last been seen when it was "here for dinner." Jim and his wife Sue have been working through their 80s on a variety of projects for the Oregon Eagle Foundation.

Jim's experience with eagles includes taking one from its rehab site in Portland to its release site in Wasco County. Jim flew the eagle there in his Piper Cub; unfortunately the eagle chewed its way out of its box shortly before landing—Jim was seen exiting the aircraft when it was still rolling down the runway, no doubt in violation of various FAA regulations. The eagle followed shortly.

Bird photography fit naturally into Jim's educational roles, and his work was of good enough quality to appear in *National Geographic*. It also appeared in Jackman and Long's iconic *The Oregon Desert* (Caxton, 1971).

After his return from Arizona to Oregon he continued to band raptors and works on nest box projects for kestrels, part of his banding program. His attitude toward predator poisoning is expressed in the rumor (spread by him) that he once acquired the location of a series of "1080" poison traps, went out and dismantled them and barely avoided prosecution owing to his reputation and good connections. —*Jim Anderson was interviewed for this history by David Vick*

R. Bruce Horsfall was the first Oregon-based artist to achieve renown. Here are his illustrations of swallows, Dipper, American Robin, and Varied Thrush, kinglets, North-ern Flicker, and Red-breasted Sapsucker. Images courtesy of the Portland Audubon Society.

Larry B. McQueen came to Oregon in the 1960s and soon became known as one of the nation's premier bird artists. A long-time resident of Eugene, his work has appeared in field guides, on magazine covers, and in many other venues. Here we see Chestnut-backed Chickadee, American Robin, and a pair of Western Tanagers.

The top image of Trumpeter Swans is by Larry McQueen; the lower image of Sandhill Cranes in the mist is by H. Jon Janosik.

H. Jon Janosik of Hubbard, Oregon, has illustrated several field guides, including the National Geographic Society's Field Guide to North American Birds. *He is active with Artists for Conservation. We feature his Tufted Puffins, Northern Pygmy-Owl, and flying Brant.*

Ram Papish grew up in Eugene, studied art at the University of Oregon, and currently lives in Toledo on the Oregon coast. He has illustrated many books and provided art for the US Fish and Wildlife Service and other entities. His Pileated Woodpeckers and Varied Thrush were both used for book covers; the Barn Owls also appeared in a book.

Becky Uhler of Eugene is well known for her poster art and finely detailed appreciation of natural subjects. Here we showcase her Lazuli Bunting and Canada Geese.

Elva Hamerstrom Paulson of Roseburg illustrated Birds of Oregon *in 2003 and is widely renowned for her visions of the natural world. Here we display an adult and young Great Horned Owl and a Poor-will feather. Shawneen Finnegan of Beaverton illustrated* Birds of Montana, *and her work has appeared in many other venues. Although she is best known for her line art, this spectacular Sharp-tailed Sandpiper is a fine example of her color work.*

Two younger artists are showcased here. Emily Poole of Eugene recently illustrated Bird Note *for Sasquatch Books, and her art appears in a line of greeting cards. We feature a Red Crossbill head, Cliff Swallow, and Spruce Grouse. Junco Bullick, a student at Portland State University, offers a Rufous Hummingbird and Acorn Woodpeckers.*

CHAPTER 13

Development and Limits of the Citizen Scientist: 1950s to the Internet

Alan L. Contreras

Audubon Field Notes • Audubon Societies and the Eugene Natural History Society • The "Birding" Movement, 1960–1990 • Local Bird Clubs • The Southern Willamette Ornithological Club • Oregon Field Ornithologists • Journal of Oregon Ornithology • Studies in Oregon Ornithology • Regional Studies

The remainder of this history focuses on the rise of birding and the activities of people whose work with birds is not part of university or agency activity. A companion book, *As the Condor Soars*, includes advances in research and conservation activity from the mid-twentieth century through 2020.

As professional ornithology was becoming a more substantial undertaking with significant university involvement and bird specialists began joining state and federal agencies, bird study by interested lay people continued to expand. Some of the information gathered by non-academic observers proved important, in particular for knowledge of range expansions, population changes, and seabird movements.

The questions that arose in the early twentieth century regarding the nature of scientific work and the question of professionalism remain alive today. In this and following chapters, we attempt to present a variety of bird study activities in a way that clarifies where they fit into this matrix of questions.

THE EXPANSION OF FIELD NOTES

The National Audubon Society had included field reports (including early Christmas Bird Count reports) in its magazine *Bird-Lore* from 1899 to 1941, at which time *Bird-Lore* became a part of the new Audubon magazine. In 1948 the National Audubon Society launched a free-standing journal called *Audubon Field Notes*, sometimes just called *AFN*. This magazine was written for the serious observer interested primarily in bird distribution and related geographic issues. This kind of material—first local records, regional studies, and the like—had been a regular feature of the professional journals such as *Auk* and *Condor* and their predecessors. The sheer volume of local data being generated by newer generations of observers, well-provided with field guides and reasonably priced optics, needed somewhere to go, and *Bird-Lore* could not contain it all.

The success of *AFN*, which became *American Birds* in 1971 and is today known as *North American Birds*, was unchallenged until Internet-based methods of data exchange became a major factor in the early 2000s (see Chapter 19). *AFN* started out as a collection of somewhat random regional reports from locations that happened to have observers interested in writing. It did not attempt to cover every state or province. By the time the journal became *American Birds*, the reporting system was essentially state-based, with some states split between natural regions.

For a while, Oregon was split into three different regions, Northern Pacific Coast (all of Oregon, Washington, and British Columbia west of the Cascade summit), Northern Rocky Mountain-Intermountain (the northern half of eastern Oregon plus eastern Washington, Idaho, and part of Montana), and what amounted to a Great Basin/central Rockies region with a few stray areas attached. For much of the late 1960s and 1970s, John B. Crowell Jr. (co-regional editor, 1965–1977) from Lake Oswego and Harry B. Nehls (co-regional editor, 1966–1977) of Portland wrote the NPC report that covered western Oregon. Thomas H. Rogers Jr. from Spokane wrote the report for the northeastern parts, with Oliver Scott of Utah writing the Great Basin segment, until the tripartite split of Oregon ended in 1974. These were the people who saw the significant increase in reporting from most of Oregon. Oregon and Washington became a single reporting region in 1989, typically using one or two editors from each state to prepare the regional reports.

From the 1950s through the 1990s, *American Birds/North American Birds* was *the* source for trends and happenings in bird distribution in North America, particularly the United States and Canada but also expanding into the Caribbean and Mexico. However, the publication was increasingly decoupled from the mission of the National Audubon Society, which had become a general-purpose conservation organization less interested in the details of the natural world and more focused on its role and image in the national political scene.

The National Audubon Society transferred *NAB* to the American Birding Association in the 1990s. It remains based at ABA today, precariously poised on the edge of two questions: What is the future of sharing bird information on paper versus online, and, more important, how is distributional bird data to be curated for its various audiences? As of this writing, these questions appear to be answered with a hybrid response: a paper journal that includes major articles and an online set of reports featuring regional expert curation. See Chapter 19 for additional discussion of the general issue of data sharing, particularly with regard to eBird.

Most of what appeared and appears in this journal is not scientific research, though some of the major articles fall in that category. What *NAB* does is provide a record and brief analysis of changes in the geographic distribution of birds, a subject largely not covered by journals such as *Auk* or *Condor* (both published by the American Ornithological Society, successor to the American Ornithologists' Union)[1] or *Wilson Bulletin*. Thus it is the best source for information on, for example, range expansion in Oregon of Bushtit, Anna's Hummingbird, Red-shouldered Hawk, and White-tailed Kite. How *NAB*'s role will change given the advent of eBird remains to be seen, but there are basic advantages to each.

American Birding Association

The American Birding Association (ABA) was established in 1969 to provide a way for recreational bird observers to communicate about and coordinate their interests. Although it had no particular connection to Oregon, its existence symbolized change in the study of birds, for it focused much of its energy on challenges in identification and on finding unusual birds or birds in unusual places. It also published site guides, first in *Birding* magazine and then as free-standing publications.

Over time it added a variety of other services and today focuses on the needs of people, particularly the relatively well-off, who engage in significant amounts of birding-oriented international travel. It also operates a well-regarded set of birding camps for young people. Its state guide series recently added a *Field Guide to the Birds of Oregon* (2018) by Dave Irons of Beaverton.

LOCAL BIRD STUDY AND ORGANIZATIONS

Interest in birds tends to build differently in different communities. This is caused by various factors, but a key question is whether a given community has a large enough core group of interested observers and whether that core group has enough energy to expand from casual observation into educational, community, and organizational work. Sometimes one or two people are the drivers of a genuine boost in local interest and productivity. This was true of Arthur Pope's invigorating push of the NOA in the 1890s, William L. Finley's refuge advocacy and public agitation in the early 1900s, the fortuitous presence of Alfred C. Shelton at the University of Oregon before 1920, and the founders of the Portland Audubon Society in the first years of the twentieth century.

Some Early Activities of the Oregon Audubon Society

Portland bird enthusiasts were involved with the Northwestern Ornithological Association in 1894 and created the John Burroughs Club in 1898. In 1901, bird observers in Astoria, Oregon's second largest city at the time, formed the Oregon Audubon Society. In 1902 the Portland group merged with Astoria's as Oregon Audubon Society. The name changed to Audubon Society of Portland in 1966 when members agreed to affiliate with the National Audubon Society, which Portland predated.

Audubon Societies often work with school groups, mainly below the high school level, to show students what birds are. Because there were no federal or state regulations in 1902 that protected wildlife, other than game species, the first major Oregon Audubon Society undertaking was passage of an Oregon Model Bird Protection Act in the 1903 Oregon legislature. It was essentially to protect non-game birds and to end the trade in bird feathers and plumes. The Oregon bill was drafted by the American Ornithologists' Union.

Before passage, the markets in Portland as well as cities across the land displayed hanging strings of robins, thrushes, meadowlarks, warblers, and even screech owls along with the usual wild ducks, geese, and upland game birds. Many of the small birds went into meat pies, like the "four and twenty blackbirds" (the plump British kind) of the nursery rhyme. And fashion called for lots of decorative bird plumage.

Miss Metcafe, record-keeper for the society, reported that Oregon Audubon Society's efforts in 1904 were directed toward educating the public and getting the new Model Bird Law enforced. Despite the Model Bird Law fashion still called for bird plumage, much of it shipped from states without protection and from local plume hunters who flouted the law, especially on the marshes of southeast Oregon.

In 1907 the society sent every lady in "Portland's Blue Book," a social directory, a set of leaflets describing how their plumes were gathered in bloody destruction of whole colonies, with fledglings left to starve in the nest. This canny move touched the women's hearts and sped a fashion change. A note from the first society board meeting after incorporation tells how, as a result of fines imposed on Portland millinery firms for selling egret plumes, William L. Finley received $46 as informant in the cases. The reward went into the society treasury. The society also obtained $150 for the wages of two state game wardens sent to patrol Klamath area nesting colonies in the 1904 season. These efforts were largely successful.

Winters were hard on the birds in the early twentieth century with deep drifted snow and prolonged freezes that on several occasions saw the Columbia River frozen across to Vancouver, Washington. In the winter of 1914–1915 the society undertook a fund-raising campaign to buy bird feed and build and distribute shelters. Around the loop at the top of the Council Crest trolley the beneficiaries were Sooty Grouse, Ruffed Grouse, and flocks of Mountain Quail.

In the 1920s with more deep-freeze winters the society ran a feed program for waterfowl at Crystal Springs Lake and the Westmoreland ponds, and the grain was supplied by Crown Flour Mills. —*Adapted and expanded from a Portland Audubon history outline by Tom McAllister with assistance from Mary Raker Bartlett.*

Early local Audubon groups also operated in the 1920s before the "chapter" era in Baker City, La Grande, Salem, and Joseph. Formal chapters formed in Salem in 1969, Corvallis in 1970, Oakridge (see below) in 1971, and the Coos Bay-North Bend area (Cape Arago Audubon) in 1976. Additional chapters have formed in many parts of the state. These organizations work on conservation issues, operate bird counts, and conduct local field trips, among other activities.

It is worth noting that the plume trade and related destruction played a significant role in the early work of Frank Chapman and especially that of Florence Merriam Bailey, whose early books and articles achieved national success.

The Southern Willamette Ornithological Club (SWOC)

By the early 1970s, most larger communities in Oregon had a chapter of the National Audubon Society. The biggest exception was Eugene, which had two unique reasons for the Audubon absence. First, the Eugene Natural History Society was established in 1941 and became a nonprofit in 1968. In practice it functioned similarly to an Audubon chapter except that it had no connection to NAS. Because the ENHS existed and functioned as an umbrella for bird study, demand for an Audubon chapter in Eugene was soft and had no real energy behind it.

Then, in 1971, activists led by Joanne Ralston in the small mountain town of Oakridge decided that they wanted an Audubon chapter. When they were assigned their "catchbasin" for purposes of membership dues apportionment, they asked for all of Lane County and NAS agreed. Thus any NAS members in the Eugene-Springfield urban area, formerly assigned to Portland, the default chapter for Oregon, became members of the Oakridge Audubon Society. This awkward arrangement continued until members voted in 1978 to change the name to Lane County Audubon Society and meet in Eugene. The Eugene Natural History Society, loosely affiliated with the University of Oregon, continues operating to this day, offering a variety of programming to its members.

The slight gap in focus of Lane County nature clubs in the early 1970s provided a space into which a bird club similar to the 1894 Northwestern Ornithological Association (NOA) emerged. The Southern Willamette Ornithological Club was established in 1974, with George A. "Chip" Jobanek and Aaron Skirvin as principal motivators. The description below is adapted from the early history by George A. Jobanek published in *SWOC Talk* 1, no. 1 (1975). The club has continued to operate informally to this day as the Eugene "Birders Night," which also uses the original name for historic reasons.

> In early August, 1974, the idea of an active bird club was expressed in several conversations between future members. It was generally agreed that there was a need for such a club. A first meeting was planned. The first meeting was held August 12, 1974. This meeting was largely organizational. With no disagreement on the need for a bird club, the discussion revolved around the goals of SWOC and the organization of meetings and officers. It was decided to keep SWOC as informal as possible, with no formal arrangement of officers. Also presented at this meeting was a tape of the Hooded Warbler, recorded near Monroe and a tape of the courtship song of the Black-headed Grosbeak, recorded in Eugene. Eleven people attended.

The second meeting was held on September 3, 1974. This meeting was begun with a short presentation of a proposed cooperative field project dealing with the Wrentit. Following this, Tom Lund presented a talk on the Purple Martin project he has been conducting at Fern Ridge. Twelve people attended.

The third meeting was October 1, 1974. This meeting was devoted entirely to preparing study skins of birds. Don Payne supervised members' efforts. Fourteen people attended.

The fourth meeting was November 4, 1974. A Passenger Pigeon specimen was unveiled to begin the meeting. This specimen, an adult female supposedly collected in New York near the end of the nineteenth century, is now in [a private] collection in Eugene. A talk on the Bluebird Trail project presented by Aaron Skirvin was the major topic of this meeting. After this, the goals of SWOC were once again discussed with emphasis on the role SWOC can play in influencing city and county land planning. Fourteen people attended.

The fifth meeting on December 2, 1974 discussed bird-banding and its possible use in local projects was discussed. Twelve people attended.

The sixth meeting was on January 6, 1975. The major part of this meeting was a discussion of the establishment of breeding bird surveys and censuses in the southern Willamette Valley. Discussion then shifted to the possibility of a newsletter (!) and the unavoidable topic of dues. Randy Floyd concluded the meeting with a presentation of the Redtailed Hawk I-5 distribution study he conducted this fall. Twelve people attended.

The newsletter *SWOC Talk* was established in mid-1975. Because there was no true statewide bird club at the time, *SWOC Talk* started picking up subscribers all over the state. It became a de facto statewide bird newsletter within two years and was renamed *Oregon Birds* in 1977, with E. G. White-Swift serving as editor for a time.

When Oregon Field Ornithologists was established in 1980, *Oregon Birds* became its journal and *SWOC Talk* was briefly included as a separate section before ceasing as a separately named entity. All back issues of *Oregon Birds* and *SWOC Talk* are now available on the Oregon Birding Association web site.

Oregon Field Ornithologists/Oregon Birding Association

Oregon Birds was the principal outlet for Oregon bird news from the late 1970s through the early 2000s. It also provided something of a bridge between birders and professionals, as it gave researchers an outlet through which to request information (e.g., on locations of peregrine and eagle nests) and it gave hobby birders a place to read about more technical issues such as changes in taxonomy and forest practices. It also provided a venue for sharing information about birds that were not well studied in Oregon, for example, offshore seabirds.

On the establishment of Oregon Field Ornithologists (OFO), founding president Merlin S. "Elzy" Eltzroth of Corvallis wrote in *Oregon Birds*:

> On the second of February, 1980, approximately 110 dauntless birders met in Eugene and organized Oregon's first statewide birding association. By way of background, the fledgling society is an outgrowth of the Southern Willamette Ornithological Club (SWOC). . . . The new group adopted Oregon Field Ornithologists (OFO) as its name, approved a set of by-laws, after spirited discussion, and established three main goals for 1980–81:
>
> 1. To continue publishing OREGON BIRDS, relieving SWOC of those duties.
>
> 2. To assume sponsorship and funding of the Oregon Bird Records Committee. This committee was established in early 1978 as an adjunct of SWOC. . . . Clarice Watson of Eugene is currently Secretary to this group, whose function is to judge past and current reports of sightings in Oregon. They meet three or four times annually to determine which birds will be entered in official state records and which won't be based on the evidence available to the committee. Most states have similar committees.
>
> 3. To hold an annual meeting somewhere in Oregon for all members.[2]

Founding Board of Directors, Oregon Field Ornithologists (now Oregon Birding Association). Back row: Terry Morgan (Portland), Jim Carlson (Eugene), Alice Parker (Roseburg), Lyn Topits (Coos Bay). Front row: Jan Krabbe (Philomath), Elzy Eltzroth (Corvallis), Dennis Rogers (Port Orford). Of note is that this board continued Oregon's long tradition, beginning with Arthur Pope, of leadership roles held by young people: Dennis Rogers was a sixteen-year-old student at Pacific High School in Curry County when he was elected to the founding board of OFO. Photo courtesy of Steve Gordon, taken at first meeting of the board, 1980.

Oregon Field Ornithologists achieved a membership of over 400 in the late 1980s. At that time there were few ways to share local data and news other than through *Oregon Birds*. By the mid-1990s the Internet had spawned the Web and started a series of changes in how bird information is transmitted that continues today. OFO membership dropped well below 300 in the early 2000s before recovering. In 2019 there were 474 members, and the journal has changed from a quarterly to a spring and fall issue. The former carries highlights from the previous year in the field, including many photos, and is printed in color. The fall issue contains more semi-technical articles. OFO changed its name in a member referendum in 2012 and became Oregon Birding Association, which was by then a more accurate reflection of what the members did and cared about.

County and Regional Studies

The period after *Birds of Oregon* (1940) was, in addition to being a time of great expansion of interest in birds by nonprofessionals, an era of significant increase in detailed local publications providing information on the status and distribution of birds. There were earlier examples of local publications—Charles Emil Bendire's work and the early Klamath regional lists fall in that category—and Shelton's 1917 paper was an excellent example. After the 1940s this kind of work became much more common and remains invaluable today.

Some regional work was published in the form of long journal articles. Among the most complete of this genre is Gordon Gullion's "Birds of the Southern Willamette Valley, Oregon" an exceptionally detailed account that occupied twenty pages of *Condor* in May 1951. This was one of the major transition monographs for Oregon because it combined data from Alfred C. Shelton's 1917 publication discussed in Chapter 8 with the author's own field data, data provided by local amateur bird observers such as Ben Pruitt and that from newly minted academics, mostly Fred Evenden. As such, it was an example of the blend area between amateur and professional bird study. Gullion noted changes in local status from that set forth in Shelton and consulted Shelton's original field notes, held by the University of Oregon Museum of Natural and Cultural History.

This paper was therefore one of the first to begin comparing the status of birds over time in a discrete region of Oregon, following on Stanley Jewett and Ira Gabrielson's *Birds of the Portland Area, Oregon*, issued as Pacific Coast Avifauna No. 19 by the Cooper Club in 1929 and more dependent on a specimen record. Shelton's baseline of 1917 was probably the best pre-1920 regional status publication for Oregon, so Gullion had a sound basis of comparison.

Many such regional ornithologies, focused on distribution and seasonality, appeared as free-standing publications. Donald Farner's *Birds of Crater Lake National Park* (Kansas, 1952) is exceptionally well-written and unsurpassed in its detail about certain species; its information about breeding Lincoln's Sparrows in Oregon was still the best available from Oregon (along with Elver Voth's from Jefferson Park) when Marshall et al.'s *Birds of Oregon* (2003) was written just over fifty years later. Farner (1915–1988) spent five summers at Crater Lake National Park as a seasonal naturalist while a professor at Washington State University. Among his many accomplishments, he was president of the American Ornithologists' Union (1973–1975) and editor of *The Auk*.

Joseph E. Evanich Jr. of Portland, who attended college at Eastern Oregon University in La Grande, produced a phenomenal amount of work in his short life. In addition to becoming an excellent bird artist, he produced *Birds of Northeast Oregon* (1982) via the Grande Ronde Bird Club and the Oregon Department of Fish and Wildlife (ODFW). This covered Union and Wallowa Counties. He managed to update this booklet in 1992, shortly before he was lost to Oregon ornithology at age thirty-three during the AIDS epidemic. This small publication was called an annotated checklist but had more detail than such lists usually do.

Joe Evanich. Photo by Alan Contreras, about 1978.

Ralph Browning worked as an avian taxonomist for the US Museum in Washington, DC, but never forgot his southern Oregon connection and in 1975 issued *Birds of Jackson County and Surrounding Areas* as No. 70 in the North American Fauna series of the US Fish and Wildlife Service. This book was something of a bridge between traditional specimen-based ornithology and newer field observations, as Browning had grown up as an active teen birder in the Rogue Valley, learning from Tom McCamant, among others. In 2019 he published a memoir, *Rogue Birder*, in which much of that time is set forth in detail.

Other examples of free-standing book-size publications on counties of Oregon include *Birds of Coos County* (OFO, 1998), *Birds of Malheur*

County (OFO, 1996), *Birder's Guide to Umatilla County* (privately published by Aaron Skirvin, June Whitten, and Jack Simons, 2011), and *Birds of Lane County* (Oregon State University Press, 2006). Range Bayer has produced a detailed Lincoln County status publication focused on migration dates.

In addition to the county-based publications, there have been some special regional publications. The largest and most detailed of these is the late Carroll D. Littlefield's *Birds of Malheur National Wildlife Refuge*, which appeared in hardcover from Oregon State University Press in 1990 and was later issued in paperback, illustrated by Susan Lindstedt. This book covered a small area in exceptional detail, with information about habitat and seasonality unsurpassed by any previous publications on eastern Oregon. An excellent example of a focused local ornithology is David M. Fix's *Birds of the Diamond Lake Ranger District*, privately issued in 1991 and a remarkably thorough work.

Other obscure but interesting works abound. Some, such as Elver Voth's superb 1963 OSU master's thesis on the vertebrates of the Jefferson Park region of the Cascades, are available only in that form, not having been published. There are also a few such works that exist only in electronic form or which are in effect gray literature, available upon request but not published. Examples of these are Hendrik Herlyn's *Birds of Benton County*, an ongoing project, and Tim Rodenkirk's comprehensive ongoing *Birds of Coos County*. Not all studies follow the arbitrary boundaries of counties or states; see, for example, Ernest Booth's *Ecological Distribution of the Birds of the Blue Mountains Region of Southeastern Washington and Northeastern Oregon* (Walla Walla College, 1952).

Finally, there are examples of micro-ornithologies. Most of these do not appear in print, living instead in personal or shared databases. However, now and then extremely local data appear in print form, for example Paul Sullivan's self-published *Birds of Rummel Street* (2016), essentially an annotated yard list from McMinnville; George Jobanek's *List of the Birds of the Walterville Area* (1973); Danae Yurgel's annotated checklist *Birds of Wortman Park* in McMinnville (1973); and Darrel Faxon's *Birds of Thornton Creek, Vol. 1* (1991), part of the *Studies in Oregon Ornithology* series issued by Range Bayer through his Gahmken Press imprint, which also issued the *Journal of Oregon Ornithology*. In some cases even a brief seasonal status has value owing to lack of other information from the location during a given period; see, for example, John Wampole's 1957–1959 summer data from the Cape Arago region.[3]

Annotated Checklists

The local checklist has been a part of Oregon bird study since A. W. Anthony, Arthur Pope, and others started gathering information from people they knew and printing it. Local bird clubs and Audubon chapters have often published "field lists" for use in keeping track of what is seen.

These were all essentially "blank" lists, with bird names and usually nothing else. As more information became easily available, observers wanted something simple that they could have in the field with basic information in it. Examples include Gerard Bertrand and J. Michael Scott's 1971 annotated checklist of Oregon birds, the first widely distributed pocket list with status information and designed for use in rainy western Oregon: it was printed on waterproof paper. This list was issued by Corvallis Audubon Society, and later versions were updated by Elzy Eltzroth and Fred Ramsey in 1979 and Eltzroth in 1987.

A semi-official "Checklist of the Birds of Oregon" was published by Harry Nehls and Tom Crabtree in *Western Birds* in 1981. This was the first formal state list issued by the Oregon Bird Records Committee (see Chapter 14). The authors noted:

> The present bird list includes 415 species documented by an identifiable photograph on file with the Oregon Bird Records Committee or a specimen deposited in a museum to substantiate their occurrence in Oregon. Also included are 12 species whose presence on the official list is based upon a sight record only.[4]

This was followed by the Oregon Bird Records Committee's publication of the "official" state list in *Oregon Birds* in 1986. The OBRC has been the de facto keeper of the list and supportive documentation since the 1980s and publishes an updated report each year in *Oregon Birds*. Other local lists have been issued for most parts of Oregon. Such lists are not in themselves scientific contributions, but they do provide an evidentiary basis for range changes.

County checklists have been published in multiple formats. In 1993, Steve Summers and Craig Miller published *Preliminary Draft: Oregon County Checklists and Maps*.[5] This was an early attempt to get a list for each of Oregon's thirty-six counties, which aimed to improve our understanding of the distribution of Oregon birds. In 2010, Chuck Gates published online checklists for each county on the East Cascades Audubon Society website. The checklists are updated as new records are discovered.

Finally, there are publications that include significant material about birds but also cover other topics. One of the most interesting and well-thought-out of these, with perhaps the longest title of any small Oregon bird publication, is Fr. Hugh Feiss's *Glad for What He Has Made: A Guide to the Trees, Shrubs, Flowers and Birds of Queen of Angels Monastery and Mount Angel Abbey* (Mount Angel Library, 1982; 2nd ed., 1990). Feiss, a Benedictine monk and a scholar of considerable repute in his field, is still an active observer today in Idaho. William Eastman Jr.'s booklet *Eating of Tree Seeds by Birds in Central Oregon* (Oregon Forest Research Center, 1960) fills a unique niche, similar in content to Johnson A. Neff's small publication on woodpeckers. There are many other examples.

Site Guides

Site guides started appearing at a national level in the 1960s and Oregon started seeing them printed in local newsletters not long after. One of the most-anticipated and well-used books of the 1970s and '80s was Fred Ramsey's *Birding Oregon* (1978), issued by the Corvallis Audubon Society and based on his series of locally issued guides to various corners of the state.

Joe Evanich issued *A Birder's Guide to Oregon* in 1990 via Portland Audubon Society. It covered more sites than Ramsey's book but generally in less detail. John Rakestraw's *Birding Oregon* (2014) is the most recent state-wide guide. These guides provided some encapsulated status information because they discussed what birds were present at what seasons.

A number of local site guides have also been issued, such as Steve Summers's *A Birder's Guide to the Klamath Basin* (1993), which also includes the California side of the basin, and Mike Patterson's *A Guide to the Birds and Other Wildlife on the Columbia River Estuary* (1998). Among the best recent examples are Aaron Skirvin et al.'s 2011 *Birder's Guide to Umatilla County* and Susan Massey and Dennis Vroman's 2003 *Guide to the Birds of the Rogue Valley*. John Fitchen's 2004 *Birding Portland and Multnomah County* serves much of the urban area.

One unique item must be mentioned: *A Birder's Guide to the Sewage Ponds of Oregon*, issued by Bill Tice in 1999, surely one of the more useful ideas for field observers. The trend in site guides is to put them online where they can be kept up-to-date and downloaded as needed; see Chapter 19.

New Statewide Identification Books

There seems to be an unslakable demand for statewide bird guides, and evidence of this is the remarkable number and variety of beautifully illustrated guides to the birds of Oregon. The first of these were the relatively modest but useful and portable Portland Audubon paperbacks *Familiar Birds of Northwest Forests, Fields and Gardens* (David Marshall, 1973) and its companion *Familiar Birds of Northwest Shores and Waters* (Harry Nehls, 1975). Some of these birds were not all that familiar, as these pocket booklets included 239 species, most illustrated in color with paintings. The first of these used a number of paintings by Bruce Horsfall, and the latter included art by Larry McQueen; both artists are featured in the center plates of this book. Art by Seattle artist Zella Schultz was also included. Art has gone out of fashion in more modern books, which all but universally use color photographs, perhaps because they are cheaper to obtain. This is a real loss to observers, as more can be done with a given work of art than with most photos. We include a selection of both color and line art in this volume. A longer discussion of Oregon bird art can be found in *As the Condor Soars*.

Among newer books are Dave Irons's 2018 Oregon volume of the ABA series, *Field Guide to the Birds of Oregon*; John Shewey and Tim Blount's 2017 *Birds of the Pacific Northwest* (Portland: Timber Press); National Geographic's pocket-sized *Field Guide to Birds of Washington and Oregon*, edited by Jonathan Alderfer (2006); the regional *Birds of the Inland Northwest and Northern Rockies*, by Harry B. Nehls, Michael E. Denny, and Dave Trochlell (Olympia: WA: R. W. Morse Co., 2008); Jeff Gilligan and Roger Burrows's *Birds of Oregon* (Auburn, WA: Lone Pine, 2003); and Dennis Rogers's status publication *Birds of Oregon: Annotated Checklist* (Cinclus, 2021).

CHAPTER 14

Oregon's Expanding Avifauna

Alan L. Contreras

More Bird Art • New Field Guides • Records Committees • Photography •
Changes to Oregon's Avifauna since 1935

> *The Rarities Committee has been described as "a group of bird-watchers*
> *known for their experience, eyesight and imagination."*
> —ALLAN R. PHILLIPS in *The Known Birds of*
> *North and Middle America*, Part 1, page xxv (1986)[1]

From the beginnings of Oregon bird observation through the publication
of Ira Gabrielson and Stanley Jewett's *Birds of Oregon* in 1940, the specimen
was the only widely accepted report of a bird's presence that was acceptable
for scientific use, and even popular guides such as Florence Merriam Bailey's *Handbook* were ultimately based on a specimen record. Bailey herself
was an early advocate for enjoying and identifying birds in the field; one
of her popular books was called *Birds Through an Opera Glass*, alluding to
the 3x or 4x mini-binoculars popular in the late nineteenth century among
opera-goers.

Bailey's books and others of the early twentieth century began using
art by such notables as John Ridgway and the exceptionally gifted Louis
Agassiz Fuertes. By the late 1920s, George Miksch Sutton's work was also
appearing, mostly in books about eastern birds. In the Northwest, the art

of Allan Brooks, a Canadian military officer, appeared in P. A. Taverner's *Birds of Western Canada* (1926) and in Ralph Hoffmann's *Birds of the Pacific States* (1927).

The generally acknowledged pivot-point after which color illustrated field guides became the norm was the publication of Roger Tory Peterson's *A Field Guide to the Birds* in 1934. This covered the eastern half of the United States and Canada. Finally, for Oregon, the Peterson *Field Guide to Western Birds* of 1941 and the uniquely useful 1957 *Audubon Western Bird Guide* by Richard Pough, with its superb Don Eckelberry illustrations of unusual Asiatic species such as Sharp-tailed Sandpiper and Black-tailed Gull, enabled mid-century observers to go into the field with some confidence.

The net effect of the new field guides was both to respond to an increasing public demand for usable field guides and to fuel that demand. With a good field guide anyone could identify most of the birds they saw. They still needed good binoculars to see many birds; these optics did not become widely available until after World War II. Once that happened, photos of birds and the previously notorious "sight record" became ordinary.

As the focus of field ornithology moved mostly away from specimen collection and more people began reporting birds to various publications, states and a few distinct locales (e.g., Vancouver Island) began establishing repositories for reports of truly unusual birds such as passerines from the eastern United States or Siberian strays. These are generally known as records committees or rarities committees and emerged from multiple organizations. The California committee met informally in 1967 and became a formal entity in 1970. It is technically a committee of Western Field Ornithologists, as is the Hawaii committee.

The Oregon Bird Records Committee

The Oregon Bird Records Committee was formed in 1978 and predated Oregon Field Ornithologists (OFO). Its general structure and operation is based on the California model. The initial members were selected by an informal group of birders, mostly from the Southern Willamette Ornithological Club (SWOC), who met in Clarice Watson's living room in

Eugene and discussed who might be a good member. The committee was bootstrapped into existence ex nihilo and started functioning immediately, gathering photographic and descriptive records.

The first members of the Oregon Bird Records Committee (OBRC, 1978) were Alan Contreras, Eugene; Tom Crabtree, Salem; Jeff Gilligan, Portland; Robert Lucas, Salem; Alan McGie, North Bend; Larry McQueen, Eugene; Harry Nehls, Portland; Eleanor Pugh, Wolf Creek; and Steve Summers, Klamath Falls. The first secretary was Clarice Watson of Eugene.

The OBRC was originally funded by grants from many of the state's Audubon chapters and bird clubs and was "adopted" as a committee of OFO when that organization formed in 1980. It operates today as a committee of the Oregon Birding Association. The committee's records and related information can be found on the web at https://oregonbirding.org/oregon-bird-records-committee/.

The OBRC consists of nine members and up to five alternates from throughout Oregon. Members and alternates are elected by the current members. Members serve staggered three-year terms, and alternates serve one-year terms. The OBRC votes on most records electronically or at an annual meeting.

The following information is adapted and expanded from Marshall et al. in *Birds of Oregon* (2003).

Changes in Oregon Bird Status

Using current nomenclature, Gabrielson and Jewett listed 338 species as occurring in Oregon. Marshall et al. listed 486 in 2003, and today the OBRC recognizes 546 as of March 2022. Many of the additions have been seabirds (see Chapter 15) and most of the rest have been vagrants or relatively rare species that do not occur often in the state. Four of the birds included in 1940 are not currently recognized as being in Oregon, including California Condor. Some of the additions, however, have clearly expanded into Oregon since mid-century, and a few are probably new arrivals. From the southeast came the Cattle Egret, originally from Africa and probably a natural invader via Brazil and the Caribbean. Also from a generally southern direction came Great-tailed Grackle, now a regular local breeder in Malheur and Jackson Counties and an occasional breeder elsewhere.

Anna's Hummingbird, White-tailed Kite, Red-shouldered Hawk, Black Phoebe, and Blue-gray Gnatcatcher moved in from California. The kite, hawk, and phoebe were known by 1935 by one or two records. Virginia's Warbler and Gray-headed Junco are known to reach the edge of their breeding range, at least occasionally, in southeastern Oregon.

Not all of the arrivals are from the south. Least Flycatcher, Northern Waterthrush (first territorial birds in 1977), Grasshopper Sparrow, Franklin's Gull (first reported in 1948), and Barred Owl have arrived from the northeast or east (the sparrow may have been present in some areas, as it is hard to discover); all but the flycatcher are regular annual breeders and the owl and gull have become common. In addition, a few species that are not easy to find in breeding season have been determined to be Oregon breeders, including Horned Grebe, Red-necked Grebe, Black Swift, and, rarely, Common Goldeneye, Greater Yellowlegs, and probably Broad-tailed Hummingbird (annually) and Virginia's Warbler (irregularly).[2]

Among the major expansions of birds that were already in the state in 1935 have been European Starling, House Finch, California Scrub Jay, Caspian Tern, White-faced Ibis, Wrentit, Acorn Woodpecker, and Great Egret. Bald Eagles and Osprey have become far more common than they were in the mid-twen-

Mountain Quail. Illustration by Barbara Gleason, 2008.

tieth century, perhaps owing in part to changes in pesticide practices and an increased number of reservoirs.

Post-breeding movements of Elegant Tern have become regular since the 1980s. Palm Warbler and Clay-colored Sparrow are now known to be regular migrants in small numbers, mainly in mid- to late fall, and Tropical Kingbirds have become regular and predictable on the coast every October–November. Even Brown Booby now occurs semi-regularly in late summer and fall.

Perhaps the most dramatic change is the hundreds of thousands of Canada and Cackling Geese that now winter in the state, mainly in the Willamette Valley but also in the Columbia basin. These birds were present in small numbers until the grass seed industry expanded after World War II. Also using the grass seed fields are increased numbers of Killdeer and Dunlin. Trumpeter Swan had been a winter bird; it was introduced as a breeding species in the 1960s and has hung on in small numbers; it is a regular but local wintering species in small numbers, mostly in the northern half of the Willamette Valley and in the lower Columbia region, expanding recently to the lake basins of south-central Oregon.

There have been withdrawals, too. Rock Sandpiper was a fairly obvious winter bird as far south as Cape Arago through the 1970s, with some further south. Today it is rare and local south of Tillamook County, with small numbers occasionally reaching Cape Arago. Upland Sandpiper nested in small numbers, mostly in Grant and Umatilla Counties, until the early 2000s, but is now apparently extirpated as a breeder. The same is true of American Redstart, which bred in Umatilla, Union, and Wallowa Counties and occasionally west to the east slope of the Cascades until the 1990s, after which it has not been found as a breeder. It has also withdrawn from southeastern Washington.[3]

The status of Yellow-billed Cuckoo in pioneer times is not clear, though it seems to have been a fairly local bird found mainly along the Snake-Columbia system. Today it is seen only as a vagrant to the state and does not breed, though there are occasional reports of birds in appropriate habitat in summer.

Sharp-tailed Grouse is one of the biggest losses, as it was a common bird in parts of eastern Oregon into the early twentieth century and remained in the state until 1967. Oregon Department of Fish and Wildlife (ODFW) has reintroduced birds to Wallowa County; it is unclear whether such introductions will succeed.

A few species have lost huge numbers but have turned around and are coming back. The most obvious of these is the Western Bluebird, brought back in part by extensive nest-box programs run in the 1970s–1990s by dedicated amateurs such as Elsie Eltzroth (Corvallis), Hubert Prescott (Portland), and Aaron Skirvin, Barbara Combs, and Al Prigge (Eugene).

For a discussion of some of these changes and others in the west, see *A Century of Avifaunal Change in Western North America*, by Joseph R. Jehl Jr., Ned K. Johnson, et al. (Camarillo, CA: Cooper Ornithological Society, 1993) and *Trends and Traditions: Avifaunal Change in Western North America*, by W. David Shuford et al. (Camarillo, CA: Western Field Ornithologists, 2018). An Oregon-focused study is Jenna Curtis's master's thesis, "Sixty Years of Avian Community Change in the Willamette Valley, Oregon" (Oregon State University, 2014).

THE YELLOW RAIL SAGA, PART 1

An example of how a bird can be present yet not known is the Yellow Rail. The Yellow Rail was known to most Oregon observers until the 1980s solely from Albert G. Prill's 1901 specimen taken near Scio. However, John E. Patterson and Sidney Carter discovered breeding birds in Klamath County in 1926, which did not become widely known until a fortuitous discovery by University of Oregon biologist Herb Wisner, who found several volumes of unpublished data, including photos, by Patterson in a used book store.[4]

John E. Patterson, a professionally trained photographer, was hired by the Bureau of Entomology, Forest Insect Investigations in Ashland, Oregon, as an entomological ranger in 1914. He was a self-taught but highly competent entomologist and expert on bark beetles. From 1921 to 1924 he oversaw the Ashland field station of the USDA Bureau of Entomology. He retired in 1950 as assistant head of the bureau's Berkeley, California, laboratory and died in Ashland, July 31, 1962.

Patterson collected eggs but also used his photography skills to produce several volumes of superb photographs of nests with eggs *in situ*, including a Yellow Rail nest at Aspen Lake, Oregon, and a rare Eastern Kingbird nest in northern California. In addition, he and Sidney Carter took notes on additional Yellow Rail nests and those of Least Bittern and more common species in southern Oregon.

One of the curious aspects about the Yellow Rail records is that Gabrielson wrote in the introductory material to *Birds of Oregon* (1940) that Pat-

John E. Patterson about 1917. Photo from USDA Forest Service, Region 6, State and Private Forestry, Forest Health Protection. Courtesy of the Boyd Wickman Collection.

terson's notes had been provided on loan to the Biological Survey and had been used in preparation of the book. The book was written ten years after Patterson and Carter photographed the rail nest and eggs, yet their records were not mentioned in the 1940 *Birds of Oregon*, which stated that Prill's record was the only one from the state. Another odd aspect of this rail tale is that W. E. Griffee, a well-known Oregon observer and egg collector, saw the Patterson photos and published a note about them in *Murrelet* (now *Northwestern Naturalist*) in 1944.[5] This note included a confirming discussion with Patterson. However, it seems that no one took it seriously enough to look for the birds.

This species, notoriously hard to detect, was probably using this region in numbers throughout the twentieth century and simply had not been refound until a visiting observer, Loren Hughes, happened to hear them near Fort Klamath in June 1982. For part two of this remarkable story, see *As the Condor Soars*, which discusses modern research on the rails in Oregon.

CHAPTER 15

Expanding Knowledge of Oregon's Seabirds

Nolan M. Clements

As the rich diversity of ecosystems and avifauna inside of Oregon's land boundaries were being documented, the Pacific Ocean to its west was simultaneously offering pioneering naturalists enticing opportunities to expand the understanding of the state's avian ecology. Oregon's first pelagic ornithological reports came primarily from early researchers and explorers. In 1834 naturalist Thomas Nuttall invited John Kirk Townsend to join Nathaniel Wyeth's second expedition westward as an ornithologist. Townsend spent two years collecting specimens of birds, both from land and sea, discovering and describing twelve new species from Oregon and the Pacific Northwest. His early collections represent the first Oregon pelagic records, though most of the seabirds were probably collected in Hawaiian waters. Following Townsend were others, such as A. R. Woodcock, George Suckley, L. M. Loomis, F. M. Bailey, and William L. Finley, who all added to the collection of specimens and written accounts of pelagic birding experiences from off or nearshore Oregon. The majority of these specimens were not collected far offshore, but rather scavenged from beaches or collected from nearshore boats. Similarly, most, if not all, observations of pelagic species were made from shore or from small boats close to land. Very little was actually known about the status of birds at sea beyond Oregon's coastline during the 1800s and the first years of the twentieth century.[1]

After these early records, others began to delve into Oregon's offshore ecosystems. Between 1929 and 1936 Ira Gabrielson conducted as many as twenty-seven offshore surveys in an attempt to learn more about Oregon's pelagic ornithology. Gabrielson was often joined by Stanley Jewett, John Carter, or John C. Braly, curator of the Depoe Bay nature museum. Gabrielson's surveys were conducted between May and September from Newport north to the mouth of the Columbia River, but most were run out of Depoe Bay between July and September. These surveys ventured only around ten miles offshore, thus it is not surprising that they lacked species commonly associated with deeper water. For instance, Gabrielson and colleagues had Black-footed Albatross on only one of the twenty-seven total trips.

Although there had been prior surveys done in Oregon waters, all had been exploratory in nature. Gabrielson was the first to conduct focused surveys that improved the knowledge of the distribution and abundance of Oregon's seabirds. Due to the lack of prior bird-focused surveys in Oregon's offshore waters, the trips made by Gabrielson and company were rewarding. Among other species, Oregon's first records of Pink-footed Shearwater, Buller's Shearwater, and Pomarine Jaeger were produced by their efforts. Gabrielson also observed what were probably Black-vented Shearwaters, but was unsuccessful in collecting a specimen, and as a result the species was placed on his hypothetical list.

Although these surveys were as complete as possible at the time, Gabrielson and Jewett were still left hypothesizing about the occurrence of birds in deeper waters. The duo outlines thirty-three potential species to be added to the state's official list in their *Birds of Oregon* (1940), nine of which were considered pelagics. As much as Gabrielson and Jewett made headway in understanding Oregon's seabirds, their larger discovery was how much was unknown about eastern Pacific pelagic birds.[2]

Others around the same time also explored Oregon's waters in an effort to better understand the distribution and diversity of resident and migratory seabirds. Reed Ferris recorded his observations and collected specimens while traveling to coastal islands to band birds or while fishing for salmon from 1932 to 1943. Ferris's travels were mainly in late summer but spanned from May to October in Lincoln and Tillamook Counties. Ferris was accompanied by Alex Walker, Stanley Jewett, Noah Richards, and several others. John C. Braly, who worked with Gabrielson, also collected several specimens of his own, but his work was never published. Most of his specimens now reside in the San Diego Museum of Natural History.

Following Gabrielson and others' initial push to gather more offshore ornithological data, there was an increase in the number of researchers, opportunistic passengers, and naturalistic fishermen venturing further into Oregon's waters discovering seabirds in the process. Around 1947, Charles Yocum recorded seabirds far off the coasts of California and Oregon while aboard a ship for non-research purposes. While tuna fishing in August of 1957, Tom McAllister reported a Long-tailed Jaeger and eight or more Black-footed Albatrosses twenty-five miles west of Newport. Richard T. Holmes recorded many species of pelagic birds in Oregon waters in April of 1958. Of Holmes's sightings, the most notable was a Laysan Albatross at Heceta Bank. Ben King reported multiple observations of Leach's and Fork-tailed Storm-Petrels ninety miles off Oregon's coast in June of 1961. Oregon's first twentieth-century Short-tailed Albatross was recorded in 1963 by Bruce Wyatt thirty-two miles west of Yachats while aboard a research vessel. These early records of pelagic bird species represent data from a time when formal pelagic bird study was quite limited and recreational pelagic bird watching was not yet even conceived.

After taking notes of birds at sea while studying the Columbia River effluent in 1963, Gerald Sanger set out in 1964 and 1965 to conduct a more systematic study of the seabirds of Oregon and Washington. His paper detailed his methods to collect data and summarized his findings. In an effort to survey in a variety of seasons, Sanger went on seven cruises between the spring of 1964 and the fall of 1965—two in spring, two in summer, two in fall, and one in winter.

Sanger's cruise stopped at 10:00 in the morning and 10:00 in the evening each day to conduct oceanographic surveys. During periods when the ship was stopped, which was roughly four hours per day, birds were identified and counted. In some reports, Sanger noted species as "several," "frequent," or "few," while in others actual counts were recorded. Sanger filled the gaps in the otherwise quantitatively consistent data with estimated numbers. Nineteen species were recorded on Sanger's research cruises, including five species of passerines, shorebirds, and waterfowl. Of those, the most abundant species was Black-footed Albatross, making up 53 percent of all birds seen during Sanger's surveys. Sanger's deep-water research represents the first ornithological surveys to truly analyze the pelagic avian ecology of Oregon waters.[3]

Following Sanger were a handful of other researchers looking for an opportunity to better understand pelagic bird distribution, abundance, and ecology. Among them, J. Michael Scott and Gerry Bertrand spent a significant amount of time conducting their doctoral research in Oregon waters. In 1968, Scott attended Oregon State University pursuing a graduate degree focusing on seabirds. He worked on a variety of projects related to coastal ornithology including the documentation of Western and Glaucous-winged Gull hybrids, as well as surveying offshore distribution of and resource partitioning in Pigeon Guillemots, Common Murres, and Brandt's and Pelagic Cormorants. In addition to near shore surveys, Scott also conducted more passive expeditions farther offshore which were not related directly to his doctoral work. Scott routinely was the only person aboard the vessel, functioning as the captain by piloting the boat, and as the ornithologist by counting the birds he was seeing.[4]

It was out of this research and Scott's incidental offshore surveys that recreational pelagic birding in Oregon was born. On August 17, 1971, Scott and Bertrand, in conjunction with the Audubon Society of Corvallis, began leading annual oceanic birding trips. These excursions became the first organized and publicly attended pelagic trips in Oregon. Boats primarily departed from the Depoe Bay harbor during the fall months when migration was at its peak. In 1975, both Scott and Bertrand left the area, but these trips continued until the mid-1980s. A Corvallis Audubon trip in late August of 1977 recorded a Laysan Albatross in addition to other expected species.

The late 1970s and 1980s mark the point in Oregon's pelagic ornithological history when recreational pelagic birding was at its early peak. Pre-

Pacific Seabird Group inaugural meeting, 1973. Standing (left to right): David Ainley, M. T. Myers, Spencer Sealy, George Divoky, E. Knoder; kneeling: Miklos Udvardy, D. Anderson, Gerald Sanger, Mike Scott, David Manuwal. Photo by Helen C. McFarland, courtesy of J. Michael Scott.

1988 United States territorial waters only extended to three miles offshore allowing foreign factory ships and trawlers to fish much closer to shore. These large fishing vessels in turn provided an opportunity for pelagic birders to easily locate large numbers of birds flocking around them. In 1988, President Ronald Reagan extended territorial waters out to twelve miles. Although this extended distance is not extremely far offshore, foreign fishing ships usually stayed farther out than twelve miles, making it difficult for single-day pelagic trips to locate them.[5]

In addition to commercial fishing vessels, pelagic trips often ventured out to seamounts as a location of congregated bird life. Seamounts are the product of tectonic tension between the Pacific and North American plates. This tension has created dramatic topography where the seafloor drops by a kilometer or more. Because of their topographical extremity, seamounts are sites of upwelling, which is caused by the interaction of sea currents and seafloor topography. This upwelling brings nutrients to the surface and in

Oregon pelagic observers enjoy the feeding swarm around a group of trawlers. Photo courtesy of Greg Gillson.

turn attracts marine life, including birds. Among these seamounts, pelagic observers frequent Heceta, Perpetua, and Stonewall Banks, which have produced five species of albatross, including White-capped and Wandering, in addition to other oddities and large numbers of expected species.

This era of early, recreational pelagic birding also marked a time when multiple individuals began promoting pelagic trips. Oregon birder Tom Crabtree organized numerous trips in the late 1970s and into the 1980s. In 1978 Crabtree and Jeff Gilligan thought it would be of interest to run a trip in late September. At that time, few pelagic trips had gone out late in the season and this trip represented one of the first modern expeditions to do so; most previous trips had been in summer, usually in August and early September. This trip, with the Portland Audubon Society, left Newport on September 30, 1978, at 8:30 in the morning. They ventured out thirty-two miles where chum slicks of popcorn and cod liver oil resulted in the first convincing photos for the state of seven South Polar Skuas, a state high count at the time. This outing also produced the first verified state record of Flesh-footed Shearwater.

This late fall trip was repeated for the next two years, producing Flesh-footed Shearwater and large numbers of South Polar Skuas each time, in addition to other expected pelagic species. Crabtree moved to Bend, Oregon, in 1981, but in 1985 was contracted for the newly formed international tour company Field Guides Inc. to lead an annual pelagic trip out of Garibaldi. In addition to out-of-state tour members, Crabtree's Field Guides trips were also open to Oregon birders. These trips ran until the fall of 1990.

Others also organized pelagic trips in the early 1980s. Jim Rogers put together several fall trips off the coast of Curry County in 1980, 1981, and 1985. One of the first Oregon spring pelagic trips was run out of Garibaldi in late April of 1985. The trip visited several commercial shrimp boats working about thirty miles west of Lincoln County, producing Short-tailed and Flesh-footed Shearwater. Additionally, Jim Carlson coordinated a dozen or more trips in the waters offshore Coos, Lane, and Lincoln Counties from 1976 through 1987. Carlson also led spring trips in partnership with the 1983 Oregon Field Ornithologist conference in Coos Bay and again in May of 1986 out of Newport.

Beginning in 1987, one to two pelagic trips have been conducted in late September each year out of Charleston as part of the Oregon Shorebird Festival. In 1989, the Portland Audubon Society, through the leadership of Jim Johnson, Nick Lethaby, and Bob O'Brien, began offering up to four trips per year. At this stage, it was a smattering of birders across the state organizing pelagic trips with various local Audubon Society chapters or independently—there was no established recreational pelagic trip industry. These early independent trips set a bar for what species were to be expected and how future trips would operate. A 1982 pelagic trip out of Charleston brought together several observers who would make significant contributions to ornithology.

While recreational pelagic birding was increasing in popularity, the science of bird distributions and abundance at sea was also developing.

A 1982 pelagic trip out of Charleston, Coos County, brought together several observers who would make significant contributions to bird study. Left to right: Ken Knittle, long-time Washington observer; Jim Rogers, who compiled the Port Orford Christmas Bird Count for decades; Dennis Rogers, now a well-known writer and guide in Costa Rica, at about eighteen; Roy Woodall, Matt Hunter, co-editor of Birds of Oregon (2003), also at about eighteen; and Doug Stotz, now a professional ornithologist at the Field Museum in Chicago. Photo courtesy of Steve Gordon.

In 1987, Oregon birder Paul Sullivan worked on foreign fishing vessels as a fishery observer under the National Marine Fisheries Service from late July through mid-October. Although this was not a strictly ornithological endeavor, Sullivan spent his off-hours watching the swarm of birds that followed the ship, taking detailed notes and recording counts in the process.[6]

During his two and a half months as an observer Sullivan recorded most of the expected fall pelagic migrants, in addition to several uncommon species, including a Laysan Albatross in early October and a group of Leach's Storm-Petrels fifty-five miles west of Newport in early August. Sullivan also recorded eighteen species of "lost-at-sea" passerines visiting the boat, many of which were thirty or more miles offshore. Of them, a Black-throated Blue Warbler, a Blackpoll Warbler, and a Palm Warbler, all recorded in the fall, represent some of the more unexpected species.

In 1989 and 1990, K. T. Briggs, G. A. Green, and M. L. Bonnell conducted intensive surveys for marine mammals and seabirds off the coasts of Oregon and Washington. Among other things, they found that seabirds were most densely concentrated over the continental shelf and less so in offshore waters. The Briggs et al. study remains one of the most complete surveys of Oregon's offshore bird populations. Similarly, Michael Force, in cooperation with the Marine Mammal Division of the National Marine Fisheries Service, conducted extensive systematic seabird surveys off the coasts of Washington and Oregon in the 1990s.

Until 1994, Oregon, unlike California and Washington, lacked an established pelagic birding infrastructure. After attending several pelagic trips with various local Audubon Societies and becoming familiar with Oregon's pelagic birds, Greg Gillson was unexpectedly asked to lead a trip out of Garibaldi in September of 1993. This invitation ultimately led to the advent of The Bird Guide Inc., a small business entity Gillson created in 1994. The Bird Guide Inc. ran its first official trip out of Newport on August 27, 1994. About twenty passengers voyaged roughly twenty miles offshore for eight hours. This inaugural trip enjoyed large numbers of Red-necked and Red Phalaropes, Black-footed Albatrosses, Fork-tailed Storm-Petrels, and Buller's Shearwaters. Following this trip, Gillson, with help from Matthew Hunter, planned trips in the 1995 season, primarily in spring and fall. Over the next twenty years Gillson and The Bird Guide ran trips through various charter companies out of Newport, Depoe Bay, and Ilwaco, Washington.

After becoming a regular attendee on Bird Guide pelagics, Tim Shelmerdine was invited by Gillson to become a full-time guide in 1996. Shelmerdine quickly became integrated into the Bird Guide's business, sometimes even running trips for Gillson. This was the era of regular organized and guided pelagic trips with eight or more trips each year from mid spring to late fall.

Compared with previous eras in Oregon's pelagic birding history, the increased frequency of trips between 1994 and 2014 resulted in multiple astonishing discoveries. In late September of 2006, 23 South Polar Skuas were recorded west of Newport by a Bird Guide trip. In another discovery, 125 Long-tailed Jaegers, once thought to be a rare species in Oregon waters, were recorded at a processing ship at Perpetua Bank in August of 2001. Scripps's Murrelets were found on several summer deep-water Bird Guide trips. Other trips enjoyed Parakeet Auklets in December of 2012, a Yellow-billed Loon in May of 2012, and seven Horned Puffins in March of 2007.

The Bird Guide did its fair share of history making. The second North American record and first Oregon record of White-capped Albatross was found on October 5, 1996, at Perpetua Bank. Five years later perhaps the same bird was found again, marking the second Oregon record. Several years later a Bird Guide pelagic trip had the second North American record and first Oregon record of Wandering Albatross west of Newport on September 13, 2008. Bird Guide recorded Oregon's first Brown Booby in October of 1998, first live Wedge-tailed Shearwater in October of 1999, and Great Shearwater in August of 2008.

In 2014, Gillson moved to San Diego, dissolving The Bird Guide. In its roughly twenty-year existence, The Bird Guide ran over 160 trips out of four harbors hosting approximately 4,000 birders, including repeat customers. Shelmerdine took over the role of providing Oregon with regular pelagic trips by creating Oregon Pelagic Tours in 2014. The changes that Shelmerdine made were minor. He reinstated an annual winter trip and added a short spring trip. With the addition of a few new trips and the societal embrace of the internet, Oregon Pelagic Tours has hosted birders from twelve countries. Today, Oregon Pelagic Tours continues to run the majority of its normally scheduled trips through the Newport Tradewinds charter company in Newport, but also organizes pelagic trips out of Charleston for the Oregon Shorebird Festival, and the occasional trip out of Garibaldi.[7]

Like Bird Guide Inc., Shelmerdine with Oregon Pelagic Tours schedules ten to eleven trips per year of varying lengths and bird targets. Throughout its history, Oregon Pelagic Tours has enjoyed Hawaiian Petrel three times,

Black-footed Albatross. Drawing by Joe Evanich.

Mottled Petrel once, and Fork-tailed, Ashy, Black, and Wilson's Storm-Petrels on a single trip in 2015. Oregon Pelagic Tours also has never missed Black-footed Albatross in its history.

It was during this modern age of pelagic birding that the preexisting listing habit of land birders made its way to the ocean. Big Year birders who would catch the earliest red-eye flight across the country to tick off another species would also spend thousands of dollars on pelagic trips and on repositioning cruises. After large tour cruise ships spend the summer season in Alaska, they need to move south to start the winter season in Latin America. These "repo" cruises are a fraction of the cost of normal cruises and offer birders a chance at pelagic species that charter boats rarely, if ever, find.

Extreme birders use these repositioning cruises to their advantage as an easy way to get to deep water. Under normal conditions, a fishing charter boat could take hours to get to deep water where *pterodroma* petrels, Leach's Storm-Petrels, Scripp's Murrelets, and other rarities could be found. A large majority of Oregon's records of *pterodroma* petrels have come from "repo" cruises. During Russ Namitz's record-breaking Oregon Big Year, he utilized these repositioning cruises, in addition to short-term stints aboard tuna vessels, to ensure he maximized his opportunities to check off uncommon and rare pelagic species.

In *Birds of Oregon* (1940), Gabrielson and Jewett hypothesized, "If someone had the time and opportunity thoroughly to work Oregon's offshore waters, undoubtedly there would be found many regular visitors and many more stragglers of which we now know nothing." After eighty years, the limited understanding of the distribution and abundance of Oregon's offshore bird species that Gabrielson and Jewett established has become vastly expanded. As a result of the effort made by passive observers, ornithologists, and pelagic tour companies alike, Gabrielson and Jewett's hypothesis was proven true.

CHAPTER 16

Atlases, Surveys, and Counts

Paul Adamus, Alan L. Contreras, Jeff Fleischer, Chuck Gates,
Carole Hallett, George A. Jobanek, and Teresa Wicks

*The Breeding Bird Atlas of 1994–1999 • The Christmas Bird Count • Raptor
Routes • Hawk Watches • County and Locality Listings*

One of the ways that we know what has changed in the status of Oregon
birds is through various surveys, counts, and similar activities. Some of
these, such as the Christmas Bird Count (CBC), have "softer" data sets owing to the nature of the survey protocol, yet by virtue of volume and longev-
ity are of value. Others are intensely focused one-time or short-term counts
that provide a good sense of the current status of birds. These activities fall
in the gray area between scientific work and general-interest activity. They
all use different protocols, some of which, such as the atlas procedures, are
strongly regimented, while CBCs, though quantitative, are limited by their
goals and traditions. That said, even less structured counts are useful when
they are the best information available. Experts on each describe them in
this chapter.

THE BREEDING BIRD ATLAS OF 1994–1999
Paul Adamus

The Oregon Breeding Bird Atlas Project (OBBA) was intended to provide the first *systematic* statewide baseline of Oregon species distribution during the breeding season. It was conducted from 1994 to 1999. In terms of number of volunteers and geographic continuity of coverage, the OBBA remains the largest systematic multi-species survey of Oregon wildlife ever conducted. In contrast to the Breeding Bird Survey (BBS) program and the Oregon 2020 project, survey efforts did not focus on point counts at pre-specified locations. Rather, coverage was contiguous across the state. That is, the state was divided into 432 equal-sized contiguous units, and over 800 volunteers with various levels of bird identification skill were directed to collectively cover as many accessible areas as possible within each unit.

No attempt was made to estimate relative or absolute abundance of any species. Participants were free to use whatever search techniques they felt were most effective in finding and confirming the breeding of the widest variety of species within the survey units. They were encouraged to record time spent within each unit and to indicate if they lived or worked in the unit or were simply visiting. Consistent with atlas projects in other states and countries,[1] volunteers used standardized evidence codes to describe breeding as possible, probable, or confirmed. Data collected concurrently from BBS routes and other surveys were included in the OBBA data set.

Results and analyses were published as an interactive, highly detailed CD rather than as a book.[2] However, most of the maps generated by the OBBA were subsequently included, by permission, in Marshall et al. (2003). As of January 2020, the atlas has been made available online at the Oregon Birding Association's website: https://oregonbirding.org/wp-content/uploads/2020/bbs/.

In contrast to other statewide breeding bird atlas projects, OBBA volunteers surveyed birds in two types of grid units. One type was in the shape of a hexagon, with each of the 432 contiguous hexagons covering an area of 245 square miles. The other grid unit was a 9.7 square mile (5x5 km) square similar to those used in atlas projects in many other states. Unlike the hexagons, the grid of squares did not cover the state contiguously. Rather, one square was located within each hexagon, usually in the southeast corner. In situations where that position did not contain at least one mile of public road, the position of the square was shifted slightly within its hexagon.

All hexagons received some degree of coverage and 68 percent achieved their individualized targets for number of breeding species detected. Field efforts documented Oregon's only (as of the time of this publication) breeding Blue Grosbeak pair, as well as providing the strongest documentation, as of the time the project concluded in 1999, of breeding Northern Mockingbird and Red-shouldered Hawk. In addition, volunteers documented the suggestive presence during breeding season of several species which had not recently (or ever) been documented to nest in Oregon: Common Goldeneye, Merlin, Boreal Owl, Yellow-billed Cuckoo, Plumbeous Vireo, Virginia's Warbler, Black-chinned Sparrow, Pine Grosbeak, and White-winged Crossbill. In all, volunteers found 275 species and confirmed nesting of 253 (92 percent) of these.

CHRISTMAS BIRD COUNTS
George A. Jobanek and Alan L. Contreras

The Christmas Bird Count is unique in the world of North American bird study because it has existed so long—well over a hundred years—with essentially the same protocol for most of that period and the same reliance on amateur observers. Although the reliance on amateurs is not ideal, in practice many excellent observers participate and the data can be compared over a very long period of time, which makes it quite useful for long-term trends. For example, only the CBC database shows the decline and recovery of Western Bluebird populations over the twentieth century. It will also soon show the slow withdrawal of Rock Sandpiper as a wintering bird in Oregon. Articles based on CBC data have been published in *Auk, Condor, Ecological Monographs, Journal of Wildlife Management, American Birds, Journal of Field Ornithology, Wilson Bulletin,* and many others.[3]

Oregon's first Christmas bird counts were the Corvallis, Eugene, and Mulino counts in December 1912—twelve years after Frank M. Chapman started the annual winter census. On Christmas day, Harriet W. Thomson found 15 species and 377 individuals in two hours at Eugene. The next day, A. J. Stover counted 17 species and 699 individuals at Corvallis. On a cold, foggy day at Mulino, Alex Walker, just having moved to Oregon from South Dakota and not yet making cheese in Tillamook, joined by Erich J. Dietrich, listed 22 species and 441 individuals.

The 1912 count was the last count at Eugene until 1942, and at Corvallis until 1962, but Alex Walker repeated his Mulino count the next year (with Donald E. Brown), then counted at Tillamook in 1915 and at Netarts in 1920. Other early counts published in *Bird-Lore* include Forest Grove (1913, not again until 1983), Salem (1916, not again until 1963), Multnomah (1917–1918), Sodaville (1918), Monmouth (1921–1923), Medford (1932, not again until 1953), Klamath Falls (1935, not again until 1948), Ashland (1939–1940, not again until 2010), Malheur National Wildlife Refuge (first count 1939, with many years missed in the 1940s and 1950s), La Grande (1941), and Klamath Basin (1944). Alex Walker repeated his Netarts count in 1935 and counted there with his son Kenneth in 1936 and 1938. Alex, Kenneth, and Peter Walker repeated the Tillamook count in 1939.

The grande dame of Oregon CBCs is the Portland count. On December 27, 1915, Tom McCamant (barely age fourteen, later a major figure in Oregon bird circles), Jack Dougherty, and William Brewster Jr. found 17 species and 432 individuals on the first Portland census. From then to the present, there have been only three years (1924, 1925, and 1932) when no Portland count was published (the 1928–1931 counts were published not in the Christmas count section, but "The Season" section of *Bird-Lore*).

In some early years there was more than one count in the Portland area. Mary Raker made several early counts, sometimes accompanied by her father, William S. Raker, active in the Portland Girl Scout movement, and her friends. The number of count participants increased markedly when the very active Oregon (later Portland) Audubon Society became the count's sponsor.

Willard A. Eliot, author of the 1923 book, *Birds of the Pacific Coast*, Earl

A. Marshall, and Harold S. Gilbert seem to have been the principal compilers. Participants on early Portland bird counts included H. T. Bohlman, Leo Simon, Arlie Seaman, Francis Twining, Clyde Keller, Ira N. Gabrielson, J. C. Braly, Ed Averill, William H. Crowell, H. M. DuBois, B. A. Thaxter, and Stanley G. Jewett, many of them prominent as officers of the Audubon Society.

Over the years Oregon has added many counts and lost some, too. Some counts, such as Coos Bay, Coquille Valley, Cottage Grove, and Santiam Pass, have come and gone and come again as enough local observers were available. In the 2019–2020 count season, Oregon observers operated forty counts, spread fairly evenly over the state except for a dense clustering in the Willamette Valley.

Some current gaps such as northern Malheur County have had counts in the past, whereas there are other areas, notably the McNary-Hermiston area, that have significant wintering bird populations but have never had a count.

Harney County's first CBC4Kids in January, 2020, was a huge success. Leader Dr. Teresa Wicks appears kneeling at bottom right. Photo by Karen Nitz, courtesy of the Harney County Library.

In January 2020, Portland Audubon Society teamed with Harney County observers to offer a "CBC4Kids" that ran for two hours and was focused on showing children in the Burns-Hines communities what birds are in their communities. It was a cold, windy, and occasionally snowy morning, but during the two-hour count, they still managed to find 24 bird species, including Golden and Bald Eagles, a Merlin, and 360 California Quail. In the week leading up to the CBC4Kids the Harney County Library hosted a variety of events in what they called their "Bird Week."

Themed activities included making bird feeders at Storytime and the chance to compete in a Lego bird-building competition. After the count there were bird-themed games, puzzles, and, of course, prizes. All participants walked away with a bird bingo bandana, field notebook, bird whistle, and specialty prize pack provided by Friends of Malheur National Wildlife Refuge. This kind of event in a rural community has the potential to interest more people in studying and protecting birds. —*Teresa Wicks contributed to this segment*

HAWK WATCHES IN FALL MIGRATION: BONNEY BUTTE
Carole Hallett

In the late 1980s through mid-1990s Steve Hoffman, founder and executive director of HawkWatch International (HWI), was seeking sites suitable for long-term standardized counts and banding studies to investigate the timing, magnitude, and distribution of raptor migration through the Pacific Northwest. In Oregon, James Berkelman, Charles E. Stock, Paul Parrinello, Carole Hallett, John Lundsten, Daniel George, Jim Johnson, Carol Cwiklinski, Bruce Casler, and others investigated Buckhorn Lookout, Hat Point Lookout, Steens Mountain, Abert Rim, Light Peak, Mount Scott, Fremont Point, Cloudcap, Fort Stevens, High Rock, Pelican Butte, Black Butte, Larch Mountain, Green Ridge, Modoc Rim, Dog Mountain, Wind Mountain, Mount Lowe, Hawk Mountain, Bonney Butte, and other sites.

Green Ridge and Bonney Butte were the only sites that hit the mark. On September 26, 1993, Jim Johnson counted eighty-four raptors migrating south past the Terrible Traverse (TT) on the east side of the ridge. This was what we'd been searching for. Jim called Steve Hoffman, who contacted me to round up some volunteers and go see what we could find. Daniel George, Jim Johnson, myself, and others observed from both the TT and the top of Bonney Butte near the site of the old fire lookout tower. Between all of us we documented a sustained flight under various weather and wind conditions that strongly suggested the site merited further evaluation.

In 1994 HWI observers Dave Scheutze and Sean O'Conner conducted daily single-observer season-long counts at Green Ridge and Bonney Butte. We found that Green Ridge yielded higher numbers on some days under certain wind conditions, but Bonney Butte was more consistent under both east and west winds. The following year (1995) began the first year of full-season two-observer (Dave Scheutze and Alison Clark) counting at Bonney Butte and the first year of banding at the site. The banding station is located about a half mile north of the observation point in a natural clearing. Carole Hallett established the station and was lead bander 1995–1999, followed by Rick Gerhardt in 2000 and Dan Sherman in 2003.

The site now counts 2,500 to 4,500 raptors per season, bands 200–300 birds and receives approximately 200+ visitors annually between late August and the end of October. Due to geography it is sometimes socked in, gray and dismal when it is sunny down below, or it can be raining in Portland and sunny on the Butte. These conditions make it difficult to predict when viewing will be good and if the long drive will be worth the effort. Differences between Bonney Butte and other HWI sites: no kestrels! Many Merlins. Lots of Golden Eagles and Bald Eagles.

HAWK WATCHES IN FALL MIGRATION: GREEN RIDGE
Chuck Gates

Hawkwatch abandoned the idea of establishing an "official" hawk monitoring station on Green Ridge, but the seeds were sown for future action. In 2004, the East Cascades Bird Conservancy (soon to become East Cascades Audubon Society [ECAS]) decided to pick up the reins and start a migrating hawk survey of its own. Several locations were sampled, and the best site turned out to be a clear-cut discovered by Kim Boddie. This site gave a relatively unobstructed 200-degree vista allowing observers to spot hawks coming from the north. The ECAS Green Ridge Hawk Watch was born.

Since 2004, ECAS has sponsored annual surveys starting on the second weekend of September and ending on the second weekend of October. For most of its existence, the migration count has been a weekend endeavor. Each weekend during the count period, volunteers were assembled to count moving raptors. Word spread and volunteers increased to the point where it was not uncommon to see twenty volunteers searching the Green Ridge skies for raptors. That popularity exists to this day.

During the survey, a volunteer (usually the "Count Coordinator") records hourly temperatures, wind speeds, and bird detections. These numbers are collected and collated at the end of the season in an annual report and summary. At the end of the season, the data is sent to Hawkcount.org and stored in their database where it is available for public access.

Over the years, many different people have served as count coordinator for this project. Chuck Gates, Khanh Tran, Karen Sharples, Steve Dougill, Kim Boddie, David Vick and Leanna Taylor have all taken a turn at the helm of this count.

Any narrative involving the Green Ridge Hawkwatch must mention Peter Low, an expatriate from England, who is an omnipresent and integral cog in the wheel of this project. His knowledge of local raptor identification and behavior is voluminous. His ability to pick out and identify the teeny-tiny specks that become migrating raptors borders on the superhuman. He is the undisputed leader of every count season regardless of who the coordinator is.

HAWK WATCHES IN FALL MIGRATION: MODOC RIM
Stewart Janes

The site, above Upper Klamath Lake (42.24.1966, -121.49.0638), has been visited once or twice a fall since 2001 between September 7 and October 8, with most visits between September 13 and 29). The watch is conducted between 9:30 a.m. and 4:00 p.m.

Steve Dougill, Stewart Janes, and Russ Namitz at the Modoc Rim hawk watch above Upper Klamath Lake. Photo courtesy of Karl Schneck.

There are few clear trends. Weak but significant trends are seen among Red-tailed Hawks (decline) and among Accipiters (increase) by calendar date. Weather is more important. Warm but not hot days are associated with largest numbers of sightings. Observers take this to mean days in which thermal production is greatest but not so hot as to produce heat stress. A modest northwest breeze is associated with the lowest numbers. They don't appear to like a tailwind coming off the lake.

HAWK WATCHES IN FALL MIGRATION: OTHER SITES

Hawk watches are conducted occasionally at other sites in Oregon, though not on as regular a schedule as those noted above. Steens Mountain East Rim and Cape Blanco have been covered casually over the years. Steens Mountain has potential, and such species as Broad-winged Hawk clearly use that route in small numbers, but its isolated location and dependency on east winds makes it hard to cover, though it has a road usable by standard vehicles.

WINTER "RAPTOR ROUTES" IN OREGON AND BEYOND
Jeff Fleischer

Citizen science has played a significant role in understanding the lives of birds at the local, state, and national levels. Projects like Christmas Bird Counts, Breeding Bird Surveys, eBird, Feeder Watches, and many others have increased our ability to understand the world of birds. For three winters prior to the summer of 2004, I set out to enumerate birds of prey that wintered in my home county of Linn in the Willamette Valley of western Oregon. Setting up four survey transects totaling several hundred miles, I spent the winters of 2001–2002 and 2002–2003 doing monthly surveys during December through February, counting hundreds of birds each month of a variety of species.

Results were posted to Oregon Birders OnLine (OBOL). Those postings attracted a lot of interest and feedback to the point that the project was expanded with the help of a small cadre of volunteers who offered to do similar surveys in their home counties throughout the southern half of the Willamette Valley. Combined efforts on thirteen total routes during the winter of 2003–2004 accomplished several things. Thousands of birds were counted that winter showing the role this area of Oregon played in supporting wintering birds of prey. The workings of the group went very smoothly with participants sending me their results, which I then posted to OBOL. I decided to expand the project statewide and encourage even more observers interested in birds of prey to get involved in a systematic effort to enumerate them.

To set up this statewide effort, I contacted all Oregon CBC coordinators to solicit their help. The interested coordinators in turn connected local observers. I also contacted a few acquaintances from my fifteen years working as a National Wildlife Refuge manager with the US Fish and Wildlife Service. East Cascades Audubon Society offered to help make the project statewide during the winter of 2004–2005.

The project started that winter with eighty survey routes around the state. After winter number seven (2010–2011), with most of Oregon covered by the project, the decision was made to expand the project outside of Oregon. New routes were added in Idaho, California, and Washington. Through the winter of 2018–2019, our fifteenth survey season, the project expanded to 312 routes and had close to 225 primary volunteers doing the survey work throughout all of Oregon and Idaho, the Columbia River portion of Washington from the mouth east to the Tri-Cities area, and six routes in California. Over fifteen years, 46,395 hours of volunteer work had been expended to complete 10,480 surveys on 720,940 miles of transects accounting for 678,056 bird observations.

For the 2019–2020 survey season, another major expansion effort got underway to bring the project into all of eastern Washington. By the end of this season I had added sixty-four new routes in Washington and thirteen more routes in Idaho, bringing the total number of routes in the project to 389 surveyed by more than 300 project volunteers.

Data collected in this project is now being provided to the Peregrine Fund to augment their efforts to construct an international database involving all of the birds of prey species from around the world. Our data will be available to future researchers delving into population dynamics of this great family of birds.

COUNTY AND LOCALITY LISTING
Paul Sullivan

In 1980 Oregon birders began to report their listing results at the end of each year to a compiler who pulled the numbers together for publication in the journal *Oregon Birds*. This was similar to what the American Birding Association was doing nationally. For each category participants report how many species they've seen, not a list of which species they've seen. The listing results were compiled by Jim Carlson for 1980, Steve Summers for 1981–1993, Jim Johnson for 1994–1998, and Jamie Simmons for 1999–2007. Paul Sullivan has been the compiler since 2008.

The results for 1980 showed that the birder with the longest list had seen 376 species in Oregon, with thirty people reporting. There were no reports from six counties. Jim Carlson wrote:

County listing can provide much needed information for distributional studies of the birds of Oregon both because birders keep better records of where sightings were made and because more time may be spent in out-of-the-way areas.[4]

By 1982, sixty-one people reported their numbers. Jeff Gilligan broke the 400 mark for an Oregon life list. Only three counties had no reports. In 1986 Dennis Rogers and Barbara Combs became the first people to record 100 species in each of Oregon's thirty-six counties, some of which would otherwise have had very poor coverage.

In 2011 Russ Namitz set a new Oregon year list record of 381 species. In 2014 Paul Sullivan became the first person to record 200 species in each of Oregon's thirty-six counties. These efforts have led Oregon birders to pursue birds in places they otherwise would not visit. At the end of 2019, 145 people participated in reporting lists. Twenty-eight people had seen 100 species in all thirty-six counties.

In addition, many birders now do a major part of their observation by bicycle or close to home in small areas, often five-mile-radius circles, a project that encourages gathering data nearby. Although only totals are published by the Oregon Birding Association, the data, often actual counts, are frequently entered into eBird or kept in another form that can be used by those interested in the birds of a particular area, such as the notes kept for decades by Portland observer Wink Gross during his routine dog walks.

CHAPTER 17

The Long Wingspan of David B. Marshall

Alan L. Contreras and David B. Marshall

David B. Marshall (1926–2011) had perhaps the greatest impact on the future of Oregon's birds of any figure in Oregon ornithology after William L. Finley, in part because of the sheer length and breadth of his commitment to studying, protecting, and writing about birds of the state. He was also the first ornithologist born in Oregon to make a significant contribution to the profession at a national level, along with his near-contemporary Gordon Gullion of Eugene. Dave's own words below tell us much of the story of how this boy from Portland progressed through a professional career that included almost every imaginable experience and contribution.

The first of three children born to Earl and Dorothy Marshall, Dave was born March 7, 1926, and grew up in Portland, where he made frequent childhood forays into the fields and woods to observe birds. His parents were supportive of his interest in nature and enabled him to meet local professional ornithologists such as Stanley Jewett, Ira Gabrielson, nature photographer/writer William L. Finley, and noted bird illustrator Bruce Horsfall.

As an eleven-year-old, Dave made his first visit to Malheur National Wildlife Refuge in 1937. In his superb *Memoirs of a Wildlife Biologist* (Audubon Society of Portland, 2008), he says that "it became a sort of dream that

I could someday work there." Along with a few boyhood friends, he developed the habit of birding northwest Oregon by bicycle after age twelve. These "birder kids" of the late 1930s crossed the Cascade Range by bicycle when they were 15, carrying camping gear, and ranged at large throughout the Portland area and the northern Willamette Valley.

Marshall joined the Army Air Force at eighteen, where he was a B-17 ball-turret gunner and flew in several missions over Europe before the war ended. He first saw his *Auk* article on the Fremont National Forest in print as a nineteen-year-old airman on leave while visiting the home of British ornithologist Robert Coombes.

After the war he worked for the Forest Service and for Crater Lake National Park before returning to Oregon State College, where he graduated with a degree in Fish and Game Management. Marshall started his career at Stillwater National Wildlife Refuge in Nevada when it was barely operational and then went to the Sacramento refuge complex, where he was assigned to show Peter Scott his first wild Ross's Goose, and as an escort for Jean Delacour.

In September 1955 his boyhood dream came true when he was transferred from Sacramento Refuge to Malheur as the wildlife management biologist. He would serve in this position for five years before transferring to the regional office where he began serving as the regional wildlife biologist at age thirty-four, a position he held for twelve years. He was also involved in the management of new refuges in Hawaii and Alaska, but is best known for his work at Malheur National Wildlife Refuge (NWR) and in the establishment of three new refuges in the Willamette Valley of Oregon.

He conducted numerous surveys and wildlife censuses on the refuge during his tenure at Malheur. His findings are still being used today to track trends in wildlife use at the refuge. In addition to these regular projects Dave was intensely involved in the Trumpeter Swan reintroduction project at Malheur. Descendants of the original swans brought from Red Rocks Lakes, Montana, by Dave still breed on the refuge. One of his photos of a pair of dancing cranes is widely recognized and was used to create the logo for the John Scharff Migratory Bird Festival.

His photos also documented early carp eradication efforts undertaken in the late 1950s and provide documentation of ecological responses following carp control efforts. His photographs and reports are used today as the refuge crafts carp research and control plans. During his tenure at Malheur Refuge, Dave was present in 1958 for the fiftieth anniversary of the creation of the refuge; he was also present at the refuge's centennial celebration in 2008. Marshall helped establish Finley, Baskett Slough, and Ankeny National Wildlife Refuges. He also helped establish the Lewis and Clark NWR on the Columbia River, added units to the Oregon Islands NWR on the Pacific Coast, and helped establish nine new refuges on the main Hawaiian Islands.

In 1973 Marshall became chief biologist for birds and mammals in the new endangered species program in Washington, DC. Marshall rewrote the listing notice for the Bald Eagle, proposing it as Threatened in some states and Endangered in others, an approach that is now common practice but at the time was considered unusual. His willingness to work with falconers helped facilitate the partnership between The Peregrine Fund and the US Fish and Wildlife Service (USFWS) that led to one of our country's greatest conservation success stories.

During this period, his work included not only technical research into endangered species, but such improbable activities as an assignment to take two Sandhill Cranes to Tokyo as a gift to the emperor of Japan, including the unexpected need to find them overnight lodging in Anchorage (they used a pilots' lounge), all of which is set forth in the *Memoirs*.

Having wearied of Washington, he returned to Portland in 1976 to become the regional endangered species coordinator. Over the next five years, he issued more than 300 biological opinions on endangered species. Many of these opinions were highly controversial, but none received serious challenges or resulted in legal actions—a testament to his ability to grasp many sides of an issue and to put the resource first at all times. Throughout his career, Marshall played the role of bridge-builder among groups and individuals with different views but shared interests in conserving America's wildlife.

Marshall retired from the USFWS after thirty-two years in 1981. Working as a consultant, he prepared Oregon's first plan for nongame wildlife in 1986, helped assemble the first sensitive species list for Oregon, co-authored *Sensitive Vertebrates of Oregon* in 1992, and prepared a status report on the Marbled Murrelet that resulted in the species being listed as Threatened in California, Oregon, and Washington. He also played a significant role in the work of the Audubon Society of Portland after his retirement.

In 1998, he began work on the other project that had been on his mind since childhood. He was fourteen years old and already studying birds when Gabrielson and Jewett's *Birds of Oregon* (1940) was published. Working with two co-editors, one hundred writers and other supporting volunteers, he organized the writing of the monumental *Birds of Oregon: A General Reference* (Oregon State University Press, 2003), which sold 1,500 copies in hardcover, an astonishing accomplishment for a relatively expensive scientific reference weighing in at over five pounds. The book continues in print in paperback and is widely considered one of the best modern state ornithologies.

Marshall joined the American Ornithologists' Union (AOU, now the American Ornithological Society, AOS) in 1946 and became an elective member in 1974.[1] He chaired the AOU Committee on Conservation from 1973 to 1975. Among his honors are the US Department of Interior's Meritorious Service Award and the National Audubon Society's Audubon Activists Award (1998). David B. Marshall died on November 22, 2011, following a prolonged illness.

Dave Marshall's Role in the Birth of National Wildlife Refuges

Serious consideration of establishing national wildlife refuges in the Willamette Valley began in the late 1950s and early 1960s. Marshall, chairman for Region 1's Land Acquisition Refuge Committee for the Bureau of Sport Fisheries and Wildlife, conducted a two-year study of land and water resources of the Willamette Valley for the purpose of implementing the Pacific Flyway Waterfowl Management Plan. In 1963, former Regional Director Paul T. Quick stated in a letter to Honorable Wayne L. Morse, United States Senator:

An important aspect of the [Pacific Flyway Waterfowl Management] plan is the acquisition of lands suitable for development and management to protect a basic breeding population of ducks and geese; control waterfowl damage to crops which occurs in the absence of suitable feeding and resting grounds; and make more adequate provision for recreational enjoyment and use of the waterfowl resource, including public hunting.

Regional Director Quick also stated, "It was determined that three to four areas aggregating between 10,000 and 13,000 acres should be acquired to accomplish the waterfowl management plan objective. At present waterfowl are concentrated at the north and south ends of the valley to a degree which seriously limits opportunities for recreational use of the resources."

Marshall identified seventeen sites in his assessments. In 1963, the region decided to pursue five of them for acquisition, and three eventually became part of the Willamette Valley National Wildlife Refuge Complex. The following short memoir adds some personal notes to Dave's story.

Dave Marshall (foreground) at William L. Finley National Wildlife Refuge, May 2009. Back row left to right: refuge biologist Jock Beall, Tom and Barbara McAllister, Georgia Leupold Marshall, and Molly Monroe. Photo courtesy of US Fish and Wildlife Service.

SEVENTY YEARS OF A YOUNG BIRDER'S EXPERIENCE IN OREGON[2]
David B. Marshall

I am now eighty-four years of age, which allows me to compare how bird study has changed over several generations. Since various family members were active Audubon members and took birding trips, numerous mentors entered my life. Among them was Stanley G. Jewett, regional biologist for the then US Biological Survey and co-author with Ira N. Gabrielson for the first *Birds of Oregon* (Oregon State College, 1940). In my youth I had only two birding friends, Tom McAllister and Bill Telfer. We came together through the then Oregon Audubon Society. We developed birding skills that went way beyond most of the prominent Audubon members, many of whom had hearing problems. As teenagers and before, most of our transportation was by bicycle, but as young people are discovering today, bicycle birding has its advantages.

Tom McAllister and I were graduates of the class of 1950 from OSU's fish and wildlife department, then called Department of Fish and Game Management. Insofar as I can remember, there were approximately forty students in our class who were fish and game management majors. With one exception, all were World War II veterans. There were no female students majoring in fish and game, and all except Tom and I entered the department through an interest in hunting and fishing rather than studying birds. We received great faculty support. Large numbers of birders entering this field today constitute another big change. The only other birder at OSU that I am aware of during this period was our friend Fred Evenden, who was seeking his PhD.

I had one other special interest, which was endangered species. With passage of the Endangered Species Act of 1973, I became involved with assessing which species met listing requirements and planning for their recovery. Perhaps no one interested me more on this subject than William L. Finley, who provided lectures and motion pictures of the California condor. This took place in the 1930s, and again I have to attribute this interest to the Oregon Audubon Society.

On Bicycle Birding

Although some parents would not permit their kids to travel to faraway places by bicycle, Daddy encouraged it despite a doubting mother. He bicycled as a boy with his brother, and felt it was well worth the risk considering what we would learn, and that it would build self-confidence and independence. He was right. Often with my brother, Tom, Bill, and I gained access to Portland birding areas via our bicycles, mostly on the east side. Favorite areas included Mount Tabor, Kelly Butte, Powell Butte, the Lynwood Road pond, wetlands now encompassed by the west end of the Portland International Airport, farmland between SE 82nd Avenue and Gresham and what was then a wild area north of Bybee Lake, that stretched along the Columbia River to Kelly Point at the mouth of the Willamette. During the summer we carried camping gear on our bicycles and rode to various places in the Cascades and into Eastern Oregon far enough to get a feel for arid areas.

Our birding was not competitive. We did not base success on who could tabulate the greatest number of species in a given area or a given time. We were interested in learning about life histories, habitat requirements, distribution, and other aspects of each species.

Early Birding

I can't say when I started birding. The family home had mammal, bird, tree, and flower books, but only the bird books got scribbled on and torn apart. The book I liked most was Willard A. Eliot's *Birds of the Pacific Coast* (G. P. Putnam's Sons, 1923). The text meant nothing to me because I couldn't read yet, but the colored reproductions of R. Bruce Horsfall's paintings taught me to identify the common local passerines and woodpeckers.

Some of my earliest memories are from the family dining room watching birds at backyard feeders serviced by my father, Earl A. Marshall, and mother, who was born Dorothy Brownell. This was on southeast 55th Avenue between Salmon and Taylor just west of Mount Tabor Park in Portland. It would have been in the late 1920s. The species composition was not the same as found there today. Often present was a covey of California Quail and an occasional Mountain Quail. There were no House Finches, Starlings, or Scrub Jays. I remember my parents having problems with Ring-necked Pheasants extracting newly planted peas from the soil. The quail have long since disappeared from this neighborhood.

It was always a treat to go to the home of my grandmother and numerous aunts and an uncle who lived in a big house on southwest Summit Drive below Council Crest in Portland, because they had Purple Finches coming to their window feeder. They fed the finches hemp seed which we can no longer get. [Maybe sometime soon. —Eds.]

I walked to school listening to Western Meadowlarks. The land across from Glencoe School on southeast Belmont between 49th and 53rd avenues was largely a field then. Some of my classmates thought my birding interests to be rather strange, but Portland schools had classes then in what they called nature study. My nature study teacher was a lady named Linda Koch, who was a staunch supporter of my interests and had me periodically get up in front of classes to tell what I had seen.

Birding was different then. Missing were the fine optical equipment and field guides we have today. Most people birded with four-power field glasses. It was the depth of the Depression. Binoculars were out of most people's reach and far more expensive than today. Spotting scopes were unheard of. There were no bird guides as we know them today. The closest thing we had to a field guide was Ralph Hoffmann's *Birds of the Pacific States* published by Houghton Mifflin in 1927. P. A. Taverner's *Birds of Western Canada* published in 1928 by the National Museum of Canada was also very useful. None of these books showed all of our birds in color.

Malheur in 1939

I was thirteen years old in spring, 1939, and about to spend a week participating in a trip that would have a profound influence on my life. Although I had an intense interest in birds, I did not until then realize I could make a living using that interest. On June 10 of that year, twenty-five Portland Auduboners (members of the then Oregon Audubon Society) departed Portland by private cars to spend a week at what was then called the Malheur Migratory Bird Refuge. My Uncle Lou and Aunt Edna, otherwise Mr.

and Mrs. C. L. Marshall, were the trip organizers. They were also the first editors of the *Audubon Warbler*.

We left Portland about 9 o'clock on a Saturday morning, taking what is now Highway 26 through Government Camp. The route to Bend from there was via Maupin, as the present highway through Warm Springs was still many years away. The hill out of the Deschutes Canyon south from Maupin proved, as expected, too much for some of the cars, which suffered boiling radiators, but we made it to the Redmond Hotel for dinner.

At the hotel we were joined by Stanley G. Jewett and his wife, Edna. From Bend the next day our party became a seven-car caravan with Jewett leading the way in a black 1937 Pontiac sedan that carried USDA license plates and Biological Survey logos on the front doors. As a youngster, I thought that was really prestigious. Little did I know of the public abuse one takes in driving a government car. At Horse Ridge, Jewett pointed out a Brewer's Sparrow, a bird I had never heard of. This is not surprising, considering there were no Peterson field guides for the West at that time, although I did later manage to acquire a copy of Ralph Hoffmann's book.

The absence of field guides was not the only handicap. I recall Jewett was the only member of the party who had binoculars; his were a government issue 7x35 Bausch & Lomb. The rest of us had to get along with 4-power field glasses, which I found to be of little help. While this lack of equipment may surprise today's readers, consider the fact we were coming out of the Great Depression, and binoculars cost over $200 at a time when the Chevrolet Suburban was purchased new for about $1,000.

We reached Burns in time to lunch on the expansive lawn surrounding the home of Dr. E. L. Hibbard, a retired dentist. Jewett asked Dr. Hibbard if he wanted to join us on the trip, and he answered in the affirmative. The caravan headed for Wrights Point looking at ducks and shorebirds en route, just as we do today. On Wrights Point I remember my first Rock Wren. It was also a first for most of the others who had not previously visited southeastern Oregon. We traveled south along the route of what is now Highway 205, then a dusty road gouged out like a ditch through the greasewood and rabbitbrush. We spent an hour or so at the Buena Vista Pond overlook

while Stanley Jewett and Dr. Hibbard gave us all a course in waterfowl identification. Virtually every duck found at Malheur was on Buena Vista Pond,
and the air was filled with Cliff Swallows—as was the case wherever there
were rimrocks or buildings.

We spent most of the week birding from dikes and roads in the Blitzen
Valley and at one point visited refuge headquarters. Everywhere there were
signs of recent construction—dikes, canals, roads, fences, and the refuge
headquarters. Birding wasn't much in the trees at headquarters then. They
were just being planted. Most of the time George M. Benson accompanied
us. He first went to work on the refuge as a warden for Malheur and Harney
Lakes long before purchase of the Blitzen Valley in 1935. At the time of our
trip, Benson lived in a house on the east edge of what we now call Benson
Pond, which lies along the east side of the Central Patrol Road about ten
miles north of the P Ranch. His house was full of mounted birds, many of
which are now displayed at the museum which bears his name at refuge
headquarters.

One day was spent on Steens Mountain. We got no further than Fish
Lake. It took a long time to get there, as rocks had to be moved from the
center of the road before some of the cars could clear them. Jewett's government Pontiac broke down and had to be left to be towed back later. One
can readily predict how today's much lower cars would do on such a road.
However, the trip to Steens Mountain was well worth it for seeing Red-
naped Sapsuckers and other aspen specialties.

I still have my bird list from that trip. There are only eighty species on
it. I obviously overlooked putting some down. We didn't see anything unusual, but many of the species were new to the Portland party. There was a
nesting pair of Peregrine Falcons on a shelf along the rim in the south part
of the Blitzen Valley. Sandhill Cranes were not nearly as numerous as today.
Missing were Franklin's Gulls which didn't begin using Malheur until later.
Neither did we see White-faced Ibis.

My interest in birds became more intense following the Malheur experience. Gabrielson and Jewett's *Birds of Oregon* (Oregon State College,
1940) was published soon thereafter. Although not an identification guide,

it was a great help to know what to expect and to use the proper nomenclature. Common names of birds then denoted the subspecies.

My earliest trips made alone specifically for birding were to Mount Tabor Park where I often went after school. One weekend morning in about 1939 I met another birder there. He had a fine German binocular, which I envied. We introduced ourselves. The birder was Norbert Leupold of the same family which the world-famous scopes are named for. I told Norbert he should join the Oregon Audubon Society. He did, and served several terms as its president.

Mrs. T. H. Brown of the Oregon Audubon Society headed up a Portland area bird record network about this time. I soon became involved with making monthly species lists with arrival and departure dates for the southeast Portland area. My work was screened by Leo F. Simon. I was first assigned the Mount Tabor area. I got around by bicycle, and as I got older extended my range to areas east of southeast 82nd Avenue. The edge of town was 82nd Avenue and beyond that were cultivated fields and pastures with Vesper Sparrows and sometimes Horned Larks. Kelly and Powell Buttes supported Sooty Grouse.

In about 1940 I met Tom McAllister, now an outdoor writer for the Oregonian.[3] A lifelong friendship started here. We were soon making bicycle birding excursions all over east Portland, to Powell Butte, Mount Scott, and sections of the Columbia River bottoms which have since been swallowed up by the Portland International Airport. It was soon a threesome with Bill Telfer, who later became a biology professor. Our folks took us to the coast, the Cascades, and eastern Oregon insofar as gas rationing would permit, but we also camped out and bicycled to such places. We could readily bicycle Highway 26 to Mount Hood, having to get onto the shoulder only occasionally to enable a car to pass.

Good Optics and Adulthood

I can't remember exactly when I finally got a reasonably good binocular, but I suspect it was about 1942. It was a surplus World War I artillery 6x30 Bausch and Lomb which my father purchased for me. Except for the lenses, it was solid brass. He paid $35.00 for it, which I suspect in actual purchasing power was equivalent to about $350.00 today. Invariably, we saw, or thought we saw, something rare. The next step was to call Jewett. He taught us to be good observers, to always question what we saw, and helped us learn how to keep notes and make accurate observations.

One of my first accepted finds were House Finches near southeast 45th and Holgate on 5 May 1941. Gabrielson's and Jewett's *Birds of Oregon* put the northern limits of the House Finch's western Oregon range as the Umpqua Valley. The year 1941 was the start of the House Finch invasion into the Portland area. European Starlings and Brown-headed Cowbirds were unknown in the Portland area before World War II. House Sparrows

seemed more conspicuous than today. Yellow Warblers frequented shade trees throughout Portland and Common Nighthawks could be heard every summer evening over downtown office buildings.

The summer of 1944 found my boyhood birding companions and me in military service. Our civilian birding experiences had ended for the time being. However, the birding didn't end.

Dave Marshall at about age twenty (in approximately 1946) with a young Red-winged Blackbird at what is now Finley National Wildlife Refuge. Photo courtesy of the Marshall family.

Wherever military service took us, we had an interest that made life away from home far more fulfilling than that of our fellow servicemen. —*Carla Burnside and Molly Monroe contributed information to this chapter.*

CHAPTER 18

The Life History of
Birds of Oregon: A General Reference

Matthew G. Hunter

The idea for *Birds of Oregon: A General Reference*, known as BOGR in casual usage, was conceived in the mind of a teenager in 1940. David Marshall was fourteen years old when he first perused the newly published *Birds of Oregon*, by Ira Gabrielson and Stanley Jewett (1940). This was the first comprehensive, in-depth, scientific treatment of birds in Oregon. David was a young and talented birder and budding biologist. He was thoroughly impressed with the tome, but thought to himself, "This will have to be re-done some day."

That seed lay dormant for nearly sixty years while Dave trekked through the demanding period of career and family. I suspect the seed for a "redo" of *Birds of Oregon* may never have seen the light of day if Dave had not been asked by Oregon State University (OSU) Press, in 1997, to review a recently submitted complete draft of a book on the birds of Oregon.

As Dave recounts it in his "Memoirs," the book submitted to OSU Press was not the type of book he had always envisioned. Based on input from multiple reviewers, OSU Press decided not to publish that book. But this event spurred Dave to create his own proposal; however, he did not rush to do this. His standards were high, and he consulted with a number of biologists, potential editors, authors, and OSU Press before preparing a thorough proposal.

As Dave was gathering his plans, ideas, and support for the book, in spring 1997 he happened to meet Alan Contreras at the Frenchglen Hotel in southeastern Oregon.[1] Dave was familiar with Alan's publications on birds in Oregon and his longtime connection with the Oregon birding community, and Alan knew of Dave's wildlife work, but they had never met. While chatting and standing in the hotel parking lot, Dave asked Alan if he would be interested in helping with the book project in a significant role. Alan said he would.

Sometime that fall, Dave and Alan were looking for a third editor to form the lead team for the book. I was one of few people in the state who was both a strong birder and a professional biologist. Alan pointed Dave to me, as I had known Alan for many years. Dave and I met, and we hit it off very well. The lead team was formed and we began meeting to further develop our plans and a proposal for OSU Press.

One component that we wanted to include in the proposal was the type and amount of artwork, and we needed to find an artist. I had a long acquaintance with artist and naturalist Elva Hamerstrom Paulson of Roseburg, daughter of ornithologists Frederick and Frances Hamerstrom, and I recommended that Dave consider her. Dave and I met with Elva and her photographer husband, Dale, at a park in Corvallis where we talked about what we wanted to accomplish and so Dave could meet the Paulsons and see some of Elva's work.

It was a great meeting. Dave could tell that Elva was the perfect person for this work, and it meant something to him that Elva was the daughter of two prominent ornithologists with whom Dave was familiar. Elva later told me that spending time with Dave reminded her of the love for wildlife that her parents had. A few of Elva's drawings were included in Dave's proposal to OSU Press, which he submitted in September 1997.

In December of 1997, Dave received a letter saying OSU Press had accepted our proposal. It was time to get down to business. As we discussed how to proceed, each editor took particular roles. Dave was the overall lead. Dave was also the primary solicitor of potential authors for species accounts, though we all brainstormed on whom to ask. Alan was the primary liaison with OSU Press, which had published other work of his. I organized and tracked the species assignments, revisions, and put most accounts on the internet for others to review. All three of us edited all species accounts and chapters of the book.

Although Dave had previously contacted a number of biologists to gauge their interest in contributing to the book, we needed the services of many dozens of biologists and birders to work on species accounts for well over 400 species. Our priority was to ask people who had worked or were currently working in some capacity with the species. We felt that this ensured we would have the most up-to-date and thorough treatment of the species possible. These folks were typically professional biologists, university researchers, or graduate students. We also solicited a few experienced birders in the state that we felt confident could do a thorough job. Ultimately, we ended up with 100 authors.

We developed guidelines for authors, including a template with specific sections or headings for each of the 476 standard species accounts. Our vision for the species accounts proved to be at least somewhat of a challenge for most authors. Professional biologists were comfortable researching scientific literature, but were often unfamiliar with birding resources and/or uneasy in incorporating "birding" knowledge into the species accounts. They also tended to be experts on a specific season, usually the breeding period. On the other hand, many birders didn't know how to find, assimilate, and extract pertinent information from the scientific literature. Nevertheless, through their perseverance and hard work, and our encouragement, we were able to produce species accounts of interest to a wide audience.

Each species account was reviewed and revised two or three times. Alan and I reviewed each first, then Dave would look at our edits, resolve any disputes, and make his own comments. When a final draft was ready, I would put each one on a website where others could review. Including all revisions, each of the editors reviewed nearly two thousand draft accounts.

Widespread use of the internet and email had come about less than a decade before beginning this project. It is undeniable that gathering and coordinating such a massive effort involving over one hundred authors would have been enormously difficult, expensive, and time-consuming without it. The use of email for communicating with all parties and the internet for posting draft species accounts saved untold amounts of time and expense. Also, since Dave was in Portland, Alan in Eugene, and I in Corvallis, email allowed us to keep in constant contact throughout the project.

We had several unique writing assignments for the book. One was writing species accounts for the 135 species on the Oregon Bird Records Committee (OBRC) review list. Dave asked Harry Nehls, secretary of OBRC for many years, if he would write these accounts, and he was happy to do so. Dave acknowledged Harry's monumental effort by naming him "Senior Contributor" for the book.

A second unique component was a treatment of taxonomy, especially subspecies. M. Ralph Browning had recently retired as taxonomist from the US Museum at the Smithsonian Institution and moved home to Oregon. We were delighted to have his assistance and this opportunity for him to flesh out his many notes and include them in the Oregon Distribution section of the book. Another important contribution was that of Rachel White Scheuering. Rachel was called upon to fill specific needs we had in order to finish the book, including additional research for some species accounts and organizing the index and literature citations.

Production of maps for all species was a task that the editors could not accomplish for this book, so we conferred with Paul Adamus about the possibility of including some, but not all, of the hexagon maps from the recently completed Oregon Breeding Bird Atlas. Paul agreed and this was a good compromise between the herculean task of complex maps for each species and a complete absence of maps. Jonathan Brooks at OSU did the work to take the OBBA data and create maps of sufficient size and clarity to use in the book, with Paul's assistance.

Throughout this process, several folks from OSU Press figured prominently in making it happen. Jeffrey Grass, director of OSU Press at the time, made sure all cogs were turning throughout the process. Jo Alexander was a constant source of help and encouragement and took the baton to the finish line when she laid out the enormous book for the printers.

The first printing of the book in 2003 was a hardbound copy with a run of 1,500 copies. These were nearly sold out within two years, and it was decided to do a second soft-bound run of 1,200 copies which came out in fall 2006. In the second printing we were able to fix a few typos and acknowledge the addition of eleven species to the state list. Book-signing events were held at Malheur Field Station, Leupold and Stevens in Beaverton (Dave, a widower, had recently married Georgia Leupold, his childhood sweetheart and recently widowed), the UO Museum of Natural and Cultural History, an Audubon Society of Portland authors event, the ABA convention in Eugene, and the OSU Bookstore.

Without doubt, beyond the book itself, one of the best things to come from the effort to produce this book was the relationships that were built or rekindled, and the acquaintances made with dozens of quality people. Most significantly for Dave, Alan, and me was the relationship we three developed through the experience. I will always remember Dave reflecting on how projects like this can sometimes challenge relationships, but we three ended up growing quite close, not only among ourselves, but our families as well.

At this writing it has now been almost nineteen years since the first edition of BOGR hit the streets. The bulk of the contents remain accurate; nevertheless, additional studies have occurred, some populations have changed (e.g., Barred Owl explosion), ranges have expanded (e.g., Black Phoebe) or contracted (e.g. White-tailed Kite), new species have been documented in the state (e.g., Zone-tailed Hawk), some taxonomy has changed (e.g., grebes after ducks), and a few names have been changed (e.g., Western to California Scrub-Jay). Is it time for a revision?

More information and detail about the project is available.[2]

Writers gather
to celebrate the
publication of
Birds of Oregon
in 2003.

Noah Strycker
and Dave Mar-
shall at the Mal-
heur Field Station
book event in
May 2003. Photo
courtesy of Bob
Keefer.

Jamie Simmons,
Hendrik Herlyn,
Harry Nehls, and
Cathy Merrifield
at the Portland
Audubon book
event. Photo
courtesy of Claire
Puchy.

Elzy and Elsie
Eltzroth

CHAPTER 19

The Internet Age

Vjera E. Thompson

As the internet became accessible, it changed how Oregon bird observers communicated, aggregated, and accessed information about birds, and connected with other birders. The changes were slow at first, then eventually became so prevalent that the former ways of communicating became mostly obsolete. The slowness of the progress can be seen in *Oregon Birds*. Before the internet, the author headers included contact information with a mailing address. In the Fall 1995 issue, an email address appeared in the President's Message header for the first time.[1] Over the next several issues, more and more email addresses were included. Email addresses started to get added to field note editors' contact information in 1996, but as late as fall 2001 were still not consistently shown.

The Rare Bird Phone Networks that were published in *Oregon Birds* were also slow to change. The concept behind the phone network was that, when a rare bird was found, the finder would call the person on the top of the phone network. That person would call two to three others, who would also call the next birders in the network, until birders across the state would be aware of the sighting. However, Oregon birder Greg Gillson remembers that the rare bird phone networks didn't work well; people often weren't home or their answering machine didn't work.[2] Eventually email replaced the function of the phone networks, but the phone network was still being published in *Oregon Birds* long after most birders had email, as late as the Spring 2000 issue.

After private emails, the first major internet communication tool was the email listserv, allowing all the listserv's subscribers to share information and discuss topics. Oregon Birders OnLine (OBOL) was Oregon's first birding listserv. Started in 1993 by a group of birders, it promoted the reporting of rare birds and discussing bird-related topics. Greg Gillson, one of the founding birders, for several years posted an email detailing OBOL's origins; here is Greg's 2002 anniversary email:

> Date: Sat Mar 2 13:23:30 2002
> Subject: OBOL anniversary
> OBOL is 9-years old [*sic*] this month!
>
> In March 1993 a Slaty-backed Gull was seen on Sauvie Island. "Oregon Birder's On Line" was formed as Rich Hoyer, Skip Russell, Bob O'Brien, Tony Mendoza, Marshall Beretta, and Greg Gillson forwarded sightings from the Portland and Corvallis areas back and forth.
>
> The first "OBOL event" occurred in November 1993 as a Scissor-tailed Flycatcher and Tricolored Heron were seen in Newport. Many of the then 60 members met each other at Newport that weekend—many met for the first time. OBOL was still being manually forwarded, primarily by Rich Hoyer and Greg Gillson.
>
> In 1994 Rich Hoyer arranged for OSU to host the list.
>
> And the rest, as they say, is history!
>
> Greg Gillson
> Cornelius, Oregon

A brief announcement of OBOL included in *Oregon Birds* in 1994 mentioned that OBOL had 70 users and was hosted by Oregon State University as of January 1994.[3] A longer announcement was published a year later in in the "Hotlines and Birding Talk by Computer" section.[4] The column goes on to explain, in detail, exactly how to send messages to a listserv and what tools and services are needed to access the internet (a phone line, dial-up subscription, and a modem were all mentioned). At this time, internet and email accounts were starting to become more widely available to Oregon birders. Rich Hoyer, one of the founders of OBOL, got an early email account as a student at Oregon State University, around 1991.[5] Dennis Arendt remembers the first time that he saw a bird that he heard about via email.[6] At the time, 1994, he didn't have email at home, but did at work. He got an email while at work informing him of the Rustic Bunting at Paul Sherrell's Eugene yard and was able to successfully chase and see the bird.

As email spread as a useful form of communication, additional listservs were started in Oregon, primarily for regional-specific areas of Oregon. During the Bundy occupation of the Malheur National Wildlife Refuge, in January 2016, an additional statewide discussion listserv was started for political and other controversial topics traditionally not allowed on OBOL. This new listserv was called Birds of Oregon (BOO).

OBOL and other listservs were instrumental in making information accessible to birders across Oregon. OBOL was intentionally designed from its beginning to include friendly discussions.[7] The fruits of this decision were evident while OBOL was still young. Bill Tice wrote an article for *Oregon Birds* which was published in 1996, three years after OBOL started. He explained how amazing OBOL was in its ability to bring together a wide variety of Oregon birders, saying:

> Imagine yourself in a large room where there are 250 Oregon birders. There are birders from all sorts of backgrounds and experience: college professors, biologists, Fish & Wildlife personnel, BLM employees, authors, and field notes editors. Some members of the Oregon Birds Record Committee are there and so is the OBRC Secretary himself! Any person can ask a question any time they care to and get an answer without interrupting other conversations.[8]

Tice goes on in the article to point out that it's not only the nature of email, but the nature of Oregon's birders that helped make OBOL a great resource: Oregon birders are willing to share with other birders.

The Greater White-fronted Goose migration illustrates Oregon birders' willingness to share. Each fall, flocks of Greater White-fronted Geese fly south through the Willamette valley high overhead, during a fairly concentrated window of time. A birder in Portland will see or hear a flock migrating overhead, and post to OBOL letting people know that a flock just flew over going south, and to be on the alert. This first email is typically followed by additional posts indicating that other observers started watching closely after receiving the first email and also were able to observe the same or additional flocks overhead.

Birding websites were the next new resource the internet provided to Oregon birders. In the same article, Bill Tice explains that not only are Oregon bird resources available on OBOL, additional resources had been collated in "what are called Web sites."[9] Several of the very first websites that were available for looking up information on Oregon birds were documented in Tice's article. These first websites were informational, with a focus on where to go birding in Oregon. They were not updated frequently but were mostly static. The website creator would put pertinent information on the site, and Oregon birders would go to the website if they needed that information.

After websites, the next technological change was web-based databases for birding records. The first birding database available in Oregon was nwbirds (later renamed to birdnotes.net), which was started by Oregon birder Don Baccus in 1999. Joel Geier, a birdnotes.net volunteer administrator, remembers that

> Don originally launched nwbirds as a "demo" project, to show . . . how easy it would be to set up an online database, using open-source software, so that volunteers could post their observations via the web. At the time he was also part of the international development team for PostgreSQL so this was a way for him to showcase its capabilities as a database engine.
>
> One further motivation that Don has mentioned came from discussions with people in Portland Audubon (Don had served on their board in the 1990s). This new database was a way to show Portland

Audubon how incidental reports by recreational birders could be harnessed for bird conservation—for example, by building species lists for public parks and other natural areas, to help inform management decisions by local governments.

This was part of the motivation for filing bird checklists and census counts for specific management areas (city parks, National Forests, etc.) rather than just by lat-long coordinates. An example of how this approach turned out to be useful was in the planning process for Luckiamute State Natural Area; Oregon State Parks' 2006 Master Plan for the area cited BirdNotes.net as the source of the bird species list given as an appendix. Several of Metro's properties in the Portland area also used BirdNotes for baseline data.[10]

A web-based rare bird alert service, the precursor to nwbirds, was announced in *Oregon Birds* 25, no. 3 (Fall 1999), as a supplement to birding email listservs. The announcement describes how Don planned to expand it from rare bird alerts to collecting sightings in the Pacific Northwest, while staying on a tight budget. When it first became functional, it was the only option available for birders who wanted to store Oregon, Washington, Idaho, and British Columbia bird records in a web-based database. By April 2002, it had expanded to accept sightings from California, Nevada, and Arizona. Utah was added later. The name and domain location were changed to birdnotes.net by 3/30/2001. Server issues in early 2021 have made the data unavailable, perhaps for good.

Both OBOL and birdnotes.net started changing the way field notes editors for *Oregon Birds* and *North American Birds* were able to gather information. Before these tools were available, information traveled by word-of-mouth, the rare bird phone network, the Portland Audubon rare bird phone mailbox, and via the postal service.

After OBOL, other regional listservs, and birdnotes.net started, editors had a lot more data available, if they knew where to look for it. Once birdnotes.net gained some users, the amount of data started to grow exponentially, making it harder for field notes editors to sort through the data. Birdnotes.net volunteer Joel Geier realized this was a problem and wrote a computer script to parse OBOL sightings into a format that would be helpful for field notes editors. The OBOL script tool was used from 2003 to 2006. Joel and Don also created quarterly reports to summarize birdnotes. net data in a format that was useful for field notes editors. These quarterly reports were published from winter 2004–2009.

The Beginnings of eBird

In 2002, Cornell Lab of Ornithology launched eBird, another online database for tracking bird sightings. It was first announced on OBOL in January 2003, and in *Oregon Birds* 34, no. 4 (November 2009). Since many Oregon birders were already using the birdnotes.net database, eBird was slow to catch on in Oregon. In 2010 eBird expanded to accept sightings from the whole world and more birders started to use it.[11]

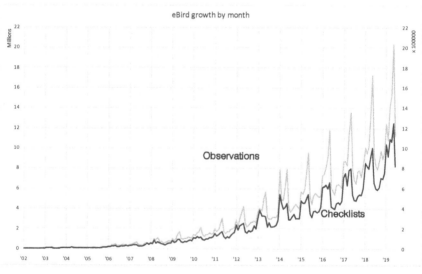

eBird growth from 2002 to 2018. Image used with permission of Cornell University.

Starting in 2012, eBird was available not only from a web browser, but also by using a smart phone application ("app"), which made it convenient to add sightings to the eBird database while out birding. Greg Gillson wrote a review of the Birdlog app on March 19, 2012, explaining how useful it was for keeping track of bird sightings, so useful that he had decided to upgrade from a "dumb" cell phone to a smart phone so that he could have access to the app.[12]

The app, now called eBird Mobile, has streamlined the ability to keep electronic notes while in the field, which in turn has changed the way that many Oregon birders and researchers collect bird data. If an organization is doing an extended survey of a location, the organizer can request that the location be added as an eBird "hotspot," which allows multiple birders and researchers to collect data and access it for later analysis. For example, the McKenzie River Trust, a nonprofit that owns and protects critical habitat in Western Oregon, set up a group of eBird hotspots in 2013 to organize surveying of one of their properties, the Waite Ranch. The hotspots have been used by trained surveyors and casual birders, with all the records available to the McKenzie River Trust, eBird users, and scientists. After the sites were added to eBird, Oregon birders added historical data as well; as of November 2019, the oldest record entered for this location was from 1987, and 157 species have been reported.

However, some are concerned that eBird data is not what it could be. Van Remsen of Louisiana State University notes that

> I can barely get my own students to write notes. eBird has let down the world of natural history by not asking for basic weather and itinerary details in the lists. Unless the list comes from someone who enters it from a smart phone, there is no way to repeat the itinerary.
>
> Without weather data and their effects on detection rates, the numbers themselves are difficult to interpret. The most important variable, wind, is not recoverable except for average wind speed at nearest weather station, which would not be a reliable indicator of local wind speed, at least in areas with any topography. And what about fog? Very local, very ephemeral. And water condition, i.e. frozen, % open water, etc. All of this is SOP for CBC data and was great training.[13]

eBird and birdnotes.net have changed the ornithological landscape by creating a way for amateurs to document birding sightings in a format that is accessible to scientists. Previously, many sightings were recorded in paper journals and are difficult to access (see Appendix E for the location of known field journals), though usually well preserved. A few of these paper journals have been uploaded to eBird, such as the twenty-nine-year data set by Karen and Jim Fairchild from their property in the Oregon coast range.[14]

One of the challenges with accepting sightings from amateurs is how to validate the sightings and catch errors. eBird has handled this by recruiting volunteer reviewers at the state or county level, who are typically long-time local birders with a solid handle on local status and distribution; as of 2019 there were twenty-three reviewers in Oregon. The volunteer reviewers maintain species filters that show expected numbers of species throughout the year. eBird software uses those filters to limit the expected species on the data entry view, and requires eBird users that "trip the filter" to document any unusual sightings or high counts with details. The reviewer analyzes the details of any sighting that is out of season or out of range and decides if the details are documented well enough to include the sighting in the scientific record. If the bird is extremely unusual, the reviewer is advised to wait until the OBRC reviews it before making a determination.

Amateur birders are not the only contributors to the eBird database. The Oregon 2020 project, a benchmark study of Oregon birds, primarily used eBird to collect data. It ran from 2013 to 2020 and surveyed a random grid of locations across Oregon in order to document Oregon's bird populations at the start of the twentieth century. The team from Oregon State University that championed the project used many internet-based tools to make the project successful: eBird to collect the data, a webpage to recruit and educate volunteers, and email to coordinate "2020 blitzes" to survey counties during the breeding season. The project used eBird to collect over 25,000 checklists, document 329 species, and survey over 2,200 locations. See *As the Condor Soars* for additional information about this project.

As these examples show, birdnotes.net and eBird changed how bird sightings were aggregated in Oregon. Before the existence of internet databases, sightings were gathered in physical notebooks, summarized in monthly Audubon newsletters that were mailed to members, and highlights were published in *Oregon Birds*. Birdnotes.net and eBird collected sightings via the internet, and made them available to anyone with access to the internet to look up reports. Sightings from multiple users were pooled together to create charts of what to expect at local birding hotspots. Email alerts also became available, with automated emails about a rare bird being sent the same day as the report.

The faster speed of the internet was both good and bad. Reports could be communicated very quickly about a rare bird, with precise details, location directions, and photos. However, a mistake or miscommunication was also distributed to many people, within seconds of sending out an email. For years there has been a tradition of a few silly birding sightings on April 1, but not all OBOL subscribers realize it's an April Fool's joke. On April 1, 2017, Tim Rodenkirk posted that he had a Magnificent Frigatebird over his place.[15] At least one birder got excited, only to discover that it wasn't an Oregon observation.

Some discussions arose on OBOL as a result of noticing possible trends, visible in almost real-time as emails were distributed via the listserv. An example of this is a project that Dave Irons initiated, using OBOL as the main form of communication, and OBOL members as the data gatherers. He started by sending this email to OBOL, on January 8, 2002:

Date: Tue Jan 8 16:50:35 2002
Subject: OBOL: Please send me your WHITE-THROATED
SPARROW reports

Obolians,
I am curious about the number of White-throated Sparrows around this winter. It seems like they are everywhere. I have found at least 9 different birds within a mile of my house and I know of a couple others nearby. If everyone could send me their Oregon White-throat records for this Winter I will compile them into some sort of unofficial grand total by county. If you have a friend who has one or more coming to feeders that

is not subscribed to OBOL please include their totals. Please list town and county for all sightings. I'm predicting the total will be in excess of 150 birds since late Fall.

I am not sure I am looking forward to all the emails but it should be fun.

Thanks,

Dave Irons
Eugene, OR

With the quick turnaround speed of email, Dave was able to quickly gather information. The next day he had heard from forty observers, and had a tally of about 125 White-throated Sparrows statewide.[16] Two days later he had heard from about fifty-five observers, and the tally had increased to 173, with reports from twenty-three of Oregon's thirty-six counties.[17] Dave's final tally was 176 White-throated Sparrows reported by fifty-nine birders, in twenty-four counties.[18]

The internet also made it possible for questions to be asked of a wide community. An example of this was in an article published in *Oregon Birds* 25, no. 1 (Spring 1999). The article was written by Elizabeth Thomas, a high school student who needed to do a science project. She decided to

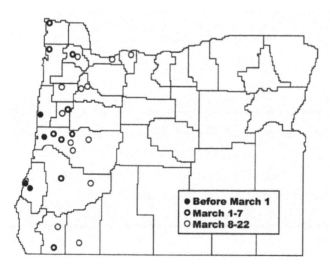

A map created by Elizabeth Thomas in 1998, showing not only the locations of Rufous Hummingbirds, but also the location of observers that Elizabeth was able to collect data from via OBOL. Courtesy of Elizabeth Thomas and Oregon Birds *journal.*

ask members of OBOL when their first Rufous Hummingbird arrived in the spring of 1997 and 1998. She was able to gather sightings from many birders across Western Oregon, create a map of the timing, and get a good grade on her science project.

The internet created ways for new connections among birders. Birders who serendipitously met in the field could recognize names from OBOL, birdnotes.net or eBird reports. In August 2016, eBird added Profile Pages which allowed birders to post a photo of themselves to their eBird account.[19] This increased the possibility that birders could recognize each other in the field even if they had never met before; in the past this would only have happened if someone's photo had been published in *Oregon Birds*. OBOL also became a tool to organize more deliberate meet-ups, such as bird walks or birding weekends.

Birders have also become more connected through birding blogs. Blogs (shortened from "web logs") are a date-oriented entry method of posting short essays or musings on a personal web page, typically longer posts with more photos than would be allowed on a listserv. This is an improvement over the early web pages available to birders, which were static, with the posted information often unchanged for long periods of time.

There are still static pages like this in use, such as the East Cascades Audubon Society's "Oregon Birding Locations by County" site guide,[20] which was first available on April 4, 2010. This website lists all the great birding locations in each Oregon county, with directions. Chuck Gates gathered the information together with the help of many birders recruited via OBOL, and data via birdnotes.net and other internet resources.[21] Although the page is designed to have details updated as site access changes, Oregon counties and birding locations will remain the main focus. In contrast, with the advent of blogs, bloggers post more frequent updates, often about local birding locations, outings, or sightings.

An example of birding blogs creating connections is the 2019 5MR challenge, encouraged by Jen Sanford on her "I used to hate birds" blog.[22] "5MR" stands for five-mile radius, so the challenge was to keep track of birds observed within five miles of one's home; to encourage driving less, and explore new areas instead of the more popular birding locations. After Jen blogged about the challenge, other bloggers also posted about it, with the result that many Oregon birders, national birders, and international birders focused on birding closer to home in 2019. The blog connection cycle continued, with other birders writing blogs about their 5MR adventures, which in turn encouraged additional local birders to visit locations described in the blogs.

With the rise of smart phones, Oregon birders now have access to vast resources in their pockets, including recent site visits on blogs. Before smart phones were available, if you wanted to have a map, directions, field guide, or recordings, you needed to carry them with you, often in a heavy or bulky format. Around 2010, David Sibley's *The Sibley Field Guide* and other field guides became available to download on smart phones, making it much easier to carry a big field guide in a compact package. As a bonus, bird recordings were also included and could be listened to in the field. In areas with internet connection, birders can also look up information they need that is posted on the internet, including recent updates on exactly where a bird was seen.

Smart phones made possible the newest iteration of Oregon birder communication methods, using WhatsApp (a centralized instant messaging service) to send short messages to a small, regionally focused group. In 2019, Portland birders created a WhatsApp group to facilitate "specifically local info that might be of less value to OBOL subscribers outside of the Portland area."[23] The concept caught on, and as of 2021, there are now seventeen different WhatsApp groups for different regional areas of Oregon. Two types of groups have emerged: a rare bird group that is used to quickly distribute information on rare birds to searchers, and an open communication group for those who want to casually share local sightings and ask questions. The open communication groups rose out of a desire to create a safe, inclusive space to welcome all birders and questions.

Smart phones and digital cameras have also made it much easier to document observations in the field. Now that cameras are included with most cell phones, photos are easier to take than previously. Birders figured out how to use binoculars and scopes (called digi-binning and digi-scoping, respectively) to provide magnification to cell phone cameras, allowing "for the record" photographs of unusual birds to be taken on the spot. In the past couple of years, cell phone audio recording equipment has become good enough that birders with a cell phone can also take recordings of birds. These photos and recordings can be added to eBird checklists, emailed to listservs, shared with a WhatsApp group, or posted on blogs, for rapid dissemination and long-term accessibility.

Smart phones also can track exactly where birders go. This has resulted in birders being able to look up good birding areas near them, to create a digital record of where they went, and to check maps to confirm where they are. eBird and other websites have provided the ability to look up nearby birding sites. eBird Mobile has also created a tracking method that saves a digital footprint connected with checklists in eBird. There are map resources that can show a birder exactly where they are, for example if they are inside their 5MR or just outside of it.

Digital photos have become a way for new observers to learn more about which birds they are seeing. They can go birding, take a picture of a bird they don't recognize, post it to a blog or discussion group, and get help identifying it long after seeing the bird. This has made birding a much more accessible hobby, in Oregon and across the world. It has also provided a new way for the Oregon birding community to hear of rare sightings.

On May 17, 2019, a photo of a Red-footed Booby was reported to the Facebook (a social media site) group called Birding Oregon. The photo was taken by a fisherman twenty miles offshore on May 16, sent to a birder who then identified and posted it to Facebook. Another birder noticed it on the Facebook group and reported it to OBOL, noting that it would be a first state record if accepted by the OBRC. The record was subsequently submitted to the OBRC and accepted.

All these new resources and technologies have changed birding in Oregon for the better. Birders can access information about Oregon's birds quickly, in the field. They can document what they saw, where, and how many, in a format that is accessible to field note editors, scientists, and other birders. There are new challenges with how to sift through the vast pile of sightings that are being created, but some wonderful new tools have already been created and yet more will be. Perhaps more important, technology has led to new generations of birders being able to document their sightings in a way that allows scientists to access them, while providing a way to connect, share, and remember Oregon bird adventures.

CHAPTER 20

Hopes, Predictions, and Expectations

Nolan M. Clements, Alan C. Contreras, Steven G. Herman,
Matthew G. Hunter, George Neavoll, M. Cathy Nowak, Mike Patterson,
Paul T. Sullivan, Vjera E. Thompson, and Teresa Wicks

The editors invited a few people familiar with bird study in Oregon to conclude this book with their thoughts about what will be happening in Oregon ornithology in the future.

Nolan M. Clements (La Grande/Corvallis, biologist and bird-bander)
As a college student, I met new people almost every week. All of them were young students working toward their dream careers. Many want to put their hands to healing, others would rather meddle in steel and concrete than bones and arteries. The question was always reciprocated, "What about you? What's your major?"

I replied that I was studying Fisheries and Wildlife Sciences, but often joked that I don't care for fish or elk, and that my real passion is for birds. Some reacted with a smile and fascinated eyes, and others quizzically asked what I plan on "doing." In almost all cases, I replied that I'd like to do research, maybe end up at a university as a faculty member, but in reality, I do not know. In ten years, when I may or may not have a doctoral degree, the avian communities will have changed. In all likelihood, more species will have gone extinct, global warming will have rendered thousands of square miles of habitat less suitable for native birds, and the plague of human expansionism will have further fragmented ecosystems.

I try not to dwell on these thoughts. Instead, I think of them as opportunities to learn, to attempt to repair, to advocate for environmental awareness and protection, and to conduct research that will make a difference. We have always had the tools that we need to make these differences—a keen eye, a sharp ear, and a thorough memory. We now have multi-thousand-dollar optics, libraries of information and pictures, and online databases. The work that needs to be done to address and solve the problems presented is not only possible, but it's currently happening in your backyard. While I tell recent acquaintances that I dream about watching birds for a living, I listen to them chuckle or see them playfully smile, and I also note the importance of bird study to the changing world that we face daily.

Alan L. Contreras (Eugene, bird observer)
I doubt the long-term superiority of electronic record-keeping. Yes, it works in the short-term and can produce some spectacular results and presentations. It is certainly convenient—I use eBird quite a lot. Yet in preparing this book I went to the University of Oregon Museum of Natural and Cultural History and read the original 1915 Lane County field notebooks of Alfred C. Shelton and those of Hubert Prescott from 1920. Everything in them was completely understandable and usable to me as a researcher looking into the status of Lane County birds over a hundred years later.

Look at how your computer software works—it is gasping for air at age ten and largely dead by age fifteen. Work I did on my computer in 1999, even in common applications, is essentially unusable in 2021 unless I made an effort to keep it current item by item. Shelton's notebooks are over five times that old, still convey exactly what they did in 1915, and will still do that in another hundred years, assuming that the paper lasts that long. By simply transcribing them onto modern acid-free archival paper we could make them, if not indestructible, remarkably long-lasting.

TR7 | Smith Rock State Park | 7/4/18
(near Terrebonne, Oregon) | 12:30–3:30 P.M.

Weather: Terrible. ~80°, full sun. Miserable on a strenuous hike like this.

With: Family, but Dad and I looked out for birds. It was certainly not "incidental," but our primary focus was managing to get to the top.

Black-billed Magpie (FOY)
(Pica hudsonia)
                        Characteristics:
← black head/ nape
← white vertical bar on wing
← glossy wing
black tail ↓
White belly →

Behavior:
Several magpies were lounging in the trees right in the parking lot! Some in the picnic area too. Somewhat secretive, but stayed still and allowed me to observe.

Species List:
(complete)
- Black-billed Magpie (5)
  (FOY)
- Common Raven (10)
- Cliff Swallow (20)
- swallow sp. (20)
- Brewer's Blackbird (5)
- House Sparrow (5)
- Red-tailed Hawk (2)
- Turkey Vulture (4)
- White-throated Swift (50) (LIFER)
- Osprey (1)
- American Robin (40)
- Northern Flicker (2)
- Buteo sp. (3)
- Canada Goose (3)
- Rock Wren (2)
  (LIFER)
- American Kestrel (1)
- Prairie Falcon (1) (FOY)
- Say's Phoebe (2)

Ashwin Sivakumar of Portland, age thirteen when these notebooks were made, keeps bird records on eBird but also the traditional way with notebooks. His work will still be usable in a hundred years. Ashwin now lives most of the time in Southern California and will be in college by the time this book appears.

The biggest advantage of electronic databases is that lots of people have records in them, and it is possible to make a fast search regarding the status of a bird. Yet speed is not always a virtue, and rushing through research on almost anything is going to produce a slipshod result. I would like to see bird observers routinely write down what they see on paper, using either a variant of the Grinnell system or something simpler. Then put them into eBird if you want to, but make sure your notebooks are preserved. In the year 2121, a young bird observer will read them in awe.

Matthew G. Hunter (Roseburg, wildlife ecologist)

I expect the stigma of "non-native species" to disappear, or at least lessen significantly. Currently, many birders and biologists despise species such as the European Starling, House Sparrow, and others. While these species do sometimes engage in acts of carnage in competition with other birds, they are not unique in this behavior, and otherwise are merely carrying out their DNA-given abilities to survive in the environment in which they find themselves.

Similarly, the Barred Owl has become the nemesis of the Spotted Owl, and the object of scorn and ridicule by biologists and birders "in-the-know." However, doesn't it seem odd that we humans have the greatest dislike for those species whose presence is largely or wholly our doing? This is the definition of the word, scape-bird. We need to accept responsibility for our own actions without despising the actions of species doing what they do to survive.

The global alteration of habitats combined with the ever-expanding mass movement of humans and their associated platforms (ships, trucks, planes) has led to more and more species being transported to places far from their evolutionary origins. Many areas near where I live are composed of as many non-native plants as native ones. Many insect populations are composed of a significant percentage of non-native species.

The "battle" to eliminate these species is lost. The habitat has changed. Being "native" is only a temporal concept. I expect a new paradigm to emerge that may include the acknowledgment of ancient natives, but will eventually simply learn to live with what is, manage environments for human values, and eliminate the need for bigoted views of certain species. As "non-native" species become "normal" and accepted parts of earth's ecosystems, we will necessarily need to manage these ecosystems to produce valuable functions and outputs that promote diversity and sustenance of life on earth, including humans.

M. Cathy Nowak (Union, biologist and bird-bander)

As I look to the future of birding, I see shadows of the past. The birders will change, as they always have. Adults will become elders, children will become adults, new birders will join the ranks. The people doing the science and those birding "just for fun" will look different on the surface: new hair styles, tattoos and piercings more commonplace, new standards for clothing appropriate for the field. But, for all these changes, the reasons for studying birds will remain the same—appreciation of their beauty and grace; fascination with their biology and life histories; wonder at their capacity for flight; amazement at long-distance migrations, and so much more.

Looking through the window of the present to the future science and study of birds, I look through reflections of the past. So much that we know, or we think we know, will have to be revisited to keep up with a changing world and birds that change with it. Advances in techniques will allow for ever finer detail in genetics and taxonomy, changing our view of relationships between species. Electronics will let us see not just where birds go but how they get there and what they encounter along the way. As climate changes, habitats will change, driving more fluid distributions and changing or eliminating migrations.

As we enter the future of ornithology and birding, I believe we will repeat much of the work of the past. We will redraw distribution maps as more species are found outside the ranges in the books we now hold. We will rewrite species accounts as we come to better understand taxonomy, biology, physiology, and behavior. We will continue to further the joy and science of birds, repeating many of the efforts of the past, as has been done for centuries. And then, in the future's future, we will do it all again.

Mike Patterson (Astoria, biologist and bird-bander)
In 1973, the last of the great lumping of bird species occurred. This "correction" came as part of a larger twentieth-century movement in biology that was based on a strict interpretation of the Biological Species Concept (BSC) and the principle of reproductive isolation. If species were known to hybridize producing viable offspring, they were lumped. Species were field-testable, reproductively discrete units based on observable morphological, behavioral, and ecological characters.

Sixteen species were reevaluated and lumped in 1973, frustrating bird-watchers, but making things more concretely manageable for working field biologists. Meanwhile, in the universities, there was a notable move to a new laboratory-based research trend that we field biologists derisively called "blender biology."

As it turned out, all that blender biology led to the unlocking of the genetic coding that is the fundamental basis of what a species is and this, in turn, precipitated a reexamination of everything we thought we knew about taxonomy. Strict interpretations of the BSC are giving way to a phylogenetic approach (PSC) based on degrees of genetic distance. Small white-faced geese are not nearly as closely related to large white-faced geese as originally thought. Falcons are more closely related to parrots than to hawks and owls. Species previously lumped are being re-split, genera are being lumped, and family trees are being re-built based on genetic relationships that are not always morphologically intuitive.

As we gain a more complete understanding of the genetic bar-code that defines them, our ability to recognize species will become increasingly dependent on what the DNA tells us in the lab. Some of the more cryptic species—species that are genetically distinct, but effectively identical in morphology—may be impossible to confidently identify without a tissue sample.

This will be the next challenge for field biologists, avocational natural historians, and educators as we work to effectively communicate the issues of species diversity and decline to a public that is increasingly skeptical of science.

The Willamette Bird Symposium at Oregon State University began in 2014. It brings together anyone interested in birds in nature. Left to right: Dr. W. Douglas Robinson, Evan Centanni, Caleb Centanni, Dr. Dan Roby, and Tye Jeske at the 2017 symposium. Photo by Alan Contreras.

Paul T. Sullivan *(McMinnville, bird observer and field trip leader)*

I expect that academic ornithologists will continue to follow their interests, using new tools to make new discoveries about the many facets of birds. I expect that amateur bird enthusiasts will continue to pursue birds in the multiplicity of ways we already have, from the casual backyard bird watcher, to the avid lister, to the aspiring citizen scientist.

All the efforts to enlist amateurs to gather data for science, for conservation, and for natural resource management will continue to be monitored for accuracy by reviewers. This will be compiled, archived, perhaps published, perhaps accessible on line, perhaps not. Notes will be lost. Publications and computers will change and no longer be accessible.

Throughout this effort, public interest, understanding, and appreciation of birds and the rest of the natural world will wax and wane with the political winds. Great efforts on behalf of birds may be set back by political leadership that cares little for nature. At other times victories will happen.

I believe victories will happen when we bring others to see the beauty of birds, to be interested in their multifaceted biology and behavior, and thus be inspired to love and protect them and all nature.

Vjera E. Thompson (Eugene, bird observer)

I started birding around 1994, about the same time my parents got their first home computer and email. OBOL connected me to other birders and made a huge difference in how quickly I was able to learn. Additional leaps in technology will impact future generations as well. Computer algorithms are being developed to analyze sounds and photos and help birders make identifications and flag data entry errors. The algorithms are not fully reliable yet, but will be in the near future. Another technology that already exists, the drone, has the potential to be used to do surveys in inaccessible locations or in unpleasant weather conditions.

The other direction that I anticipate changes is by connecting more people to birding and to each together. In the area of technology, I can foresee a smart phone app that would connect you with another birder, who could listen to the bird singing in front of you and help you identify the bird immediately. The wonderful part of this would be that technology, which often isolates us, could instead be used to connect birders.

I also hope that we can tap into the passion for micro-volunteering and create projects where many volunteers can do small tasks together to accomplish a bigger goal, like transcribing pages of scanned paper journals into an electronic database. Finally, I expect that we will see more in-person learning events, like the Willamette Valley Bird Symposium, that will allow people to learn the newest research by listening to presentations from scientists and birders.

Teresa Wicks (Burns, biologist)

As with many things regarding conservation and biology, there seems to be a common belief that technology will help "save birds." While it may be true that drones, eBird, and other technologies hold all of the answers to the future of bird conservation and ornithology, the more disconnected we become from places, the land, and the actuality of feathered bodies, the easier it is to think of birds in the abstract. Or as something removed from where we live. To address future issues associated with climate change, an increasing human population, and other sustainability-related issues, it is important to reconnect people to places, including urban and built landscapes, and the birds and plants in those places.

By connecting Oregonians, particularly young, diverse Oregonians, to places through birding, we may be able to provide better protections for birds, and create more climate-resilient landscapes. I think this needs to include connecting birders to the past through changes in bird communities, habitat loss/change, and the histories of places, particularly Indigenous management of those places. We also need to connect birders to the future of bird communities, populations, and habitats, particularly the predicted changes in bird communities associated with climate change.

The maintenance of oral and written histories of birds (i.e., field notes) will be helpful, particularly because technology can be fickle. Though some of us are more likely to keep our field notes in our bird ID books and/or on random receipts found in our pockets (guilty!).

Ultimately, our world is changing, and faster than scientists predicted. The most positive way forward is to work together to create diverse bases of knowledge and to include diverse voices at the conservation decision-making table.

Steven G. Herman, biology professor
In 2002, the *Journal of Wildlife Management* published an invited essay on the past, present, and future of the wildlife management profession by Steve Herman from Evergreen State University in Washington.[1] Because of Herman's long professional connection with Malheur National Wildlife Refuge, the Malheur Field Station, and bird study in Oregon, this was widely read in the state. Before his death in 2020 he gave permission for the editors to extract an appropriate segment for this book. —*Eds.*

In this essay, Herman noted:

We would benefit not only from reuniting wildlife biology and natural history, but . . . we might want to resurrect some of the older terms, concepts, and goals. "Management" has become a catch-all, a cover for some things that may not be at all respectable. . . . Shouldn't we bring back the term "Wildlife Conservation" and admit that preservation and sustainability are legitimate objectives in the wildlife field?

■ ■ ■

In closing this volume, let's set aside the marvels of modern technology to hear from an Oregon bird observer. George Neavoll began observing birds in Fairview, near Lebanon, Oregon, and started his life list with a California Quail at age eleven in 1949. He captures the reason why many enjoy observing birds. In a note to the editors, George said:

> Birding for me has been almost a lifesaving experience at times. We all have dark moments, but I have the memory of a Common Loon yodeling on a northern lake, walking 11 miles to see a Little Stint on the New Hampshire seacoast—and not finding it—and being in the middle of a feeding frenzy of Sooty Shearwaters at the mouth of the Columbia. Remembering such is restorative, and keeps me focused on what's important.

APPENDIX A

Oregon Ornithology Timeline

Vjera E. Thompson

Before written history	Native relationships with birds
1700s	First tentative explorations by Spanish, British and Russian ships
1804–1806	Lewis and Clark Expedition
1810–1855	Settler caravans
1825	David Douglas arrives in Oregon
1834–1836	John Kirk Townsend in Oregon
1839	Publication of John Kirk Townsend's *Narrative of a Journey Across the Rocky Mountains to the Columbia River*
1840	Thomas Nuttall's *The Land Birds* ... issued
1855–1858	Pacific railroad surveys
1859	Oregon Statehood
1875	Charles E. Bendire's first Camp Harney list published

1879	Mearns/Merrill/Wittich Klamath lists published
1880	O.B. Johnson's list for nw Oregon published
1886	Anthony's list for Washington County published
1894	Northwestern Ornithological Association founded Arthur Pope's Oregon list published
1898	Burroughs Club of Portland founded
1899	Loye Miller begins work in John Day fossil beds
1901	Lord's *Birds of Oregon and Washington* first edition published Oregon Audubon Society formed in Astoria Annie Alexander conducts work at Fossil Lake
1902	A. R. Woodcock's bird list for Oregon published by Oregon State College Oregon Audubon Society merged with Burroughs Club, kept Audubon name Florence Merriam Bailey's *Handbook of Birds of the Western United States* first published
1905	National Association of Audubon Societies established
1907	Three Arch Rocks National Wildlife Refuge established by President Roosevelt
1908	Malheur National Wildlife Refuge established by President Roosevelt Lower Klamath National Wildlife Refuge established by President Roosevelt
1909	Cold Springs National Wildlife Refuge established Deer Flat National Wildlife Refuge established
1911	William L. Finley named Oregon Game Warden
1912	First Oregon Christmas Bird Counts
1917	Alfred C. Shelton's "West-central Oregon" list published by the University of Oregon

1923   Willard A. Eliot's *Birds of the Pacific Coast* published

1927   Hoffmann's *Birds of the Pacific States* published
Schoolchildren choose Western Meadowlark as the Oregon state bird[1]
McKay Creek NWR established

1928   Johnson A. Neff's Oregon State College thesis on Oregon woodpecker diet published
Upper Klamath National Wildlife Refuge established

1929   Jewett and Gabrielson's *Birds of the Portland Area* issued by Cooper Club

1935   Oregon Islands National Wildlife Refuge established

1936   Vernon Bailey's *Mammals and Life Zones of Oregon* published
Hart Mountain National Antelope Refuge established

1938   Cape Meares National Wildlife Refuge established

1940   Landmark *Birds of Oregon* by Gabrielson and Jewett published by Oregon State College
Charles Quaintance begins teaching at Eastern Oregon University

1941   Eugene Natural History Society established

1944   Summer Lake Wildlife Area (Oregon Department of Fish and Wildlife) established

1947   Sauvie Island Wildlife Area (Oregon Department of Fish and Wildlife) established

1948   *Audubon Field Notes* begins publication, with regular Oregon reports

1949   Ladd Marsh Wildlife Area (Oregon Department of Fish and Wildlife) established

1   The choices were Western Bluebird, Varied Thrush, Oregon Junco, White-crowned Sparrow, and Western Meadowlark. With 80,000 votes cast, meadowlark won a wide majority. It was a time when you could still hear meadowlarks sing throughout the Willamette Valley and in the vacant fields while walking to school in Portland.—*Tom McAllister*

1950   E. E. Wilson Wildlife Area (Oregon Department of Fish and
              Wildlife) established

1951   Gordon Gullion publishes his work on birds of the southern
              Willamette Valley in *Condor*

1953   Wenaha and White River Wildlife Areas (Oregon Department of
              Fish and Wildlife) established

1954   Denman Wildlife Area (Oregon Department of Fish and Wildlife)
              established

1955   David Marshall named biologist at Malheur National Wildlife
              Refuge

1957   Fern Ridge Wildlife Area (Oregon Department of Fish and Wildlife)
              established

1958   Klamath Wildlife Area (Oregon Department of Fish and Wildlife,
              Miller Island) established

1959   Donald Farner, author of *Birds of Crater Lake National Park*, named
              editor of *The Auk* (1959–1963)

1961   Bridge Creek Wildlife Area (Oregon Department of Fish and
              Wildlife) established

1964   William L. Finley National Wildlife Refuge established

1965   Baskett Slough National Wildlife Refuge established
              Ankeny National Wildlife Refuge established

1966   Audubon Society of Portland agrees to become a chapter of National
              Audubon Society

1969   Jewell Meadows Wildlife Area (Oregon Department of Fish and
              Wildlife) established
              Umatilla National Wildlife Refuge established
              Salem Audubon Society established

1970   Corvallis Audubon Society established

1971 Elkhorn, Willow Creek, and Irrigon Wildlife Areas (Oregon Department of Fish and Wildlife) established
Lane County Audubon Society established as Oakridge Audubon Society
Bertrand and Scott's modern Oregon field checklist issued in Corvallis

1972 P. W. Schneider Wildlife Area (Oregon Department of Fish and Wildlife) established
Lewis and Clark NWR established

1973 Donald Farner, author of *Birds of Crater Lake National Park*, elected president of American Ornithologists' Union (AOU)
Passage of the Endangered Species Act

1974 Southern Willamette Ornithological Club established in Eugene

1975 M. Ralph Browning's *Birds of Jackson County* issued by the Fish and Wildlife Service
Coyote Springs Wildlife Area (Oregon Department of Fish and Wildlife) established
First issue of *SWOC Talk,* later *Oregon Birds*

1976 Riverside Wildlife Area (Oregon Department of Fish and Wildlife) established

1977 John A. Wiens from Oregon State University named editor of *The Auk* (1977–1984)

1978 Oregon Bird Records Committee established

1980 Oregon Field Ornithologists, later Oregon Birding Association (2012) incorporated as an outgrowth of the Southern Willamette Ornithological Club (SWOC)

1983 Bandon Marsh National Wildlife Refuge established
Lower Deschutes Wildlife Area (Oregon Department of Fish and Wildlife) established

1990 Spotted Owl listed under the Endangered Species Act

1991 Nestucca Bay National Wildlife Refuge established
Siletz Bay National Wildlife Refuge established

1992	Marbled Murrelet listed under the Endangered Species Act
	Tualatin National Wildlife Refuge established

1992
Marbled Murrelet listed under the Endangered Species Act
Tualatin National Wildlife Refuge established

1993
Oregon Birders OnLine (OBOL) begins
Mt. Bailey "renamed" in honor of Florence and Vernon Bailey
Pacific coast population of Snowy Plover listed as threatened

1994
The five-year Oregon Breeding Bird Atlas project begins

1997
George Jobanek's *Annotated Bibliography of Oregon Bird Literature Before 1935* published by Oregon State University Press

1998
Work begun on Marshall et al.'s *Birds of Oregon*

1999
Oregon Breeding Bird Atlas published as a DVD
Nwbirds, later birdnotes.net, an electronic depository for bird records, begins operation

2002
eBird, an electronic depository for bird records, begins operation

2003
Landmark *Birds of Oregon: A General Reference* by Marshall et al. published by Oregon State University Press

2009
Michael T. Murphy, Portland State University, named editor of *Auk* (2009–2013)

2012
Susan Haig from Oregon State University named president of AOU
Oregon Field Ornithologists changed name to Oregon Birding Association
Bird specimen data becomes available via VertNet

2013
Coquille Valley Wildlife Area (Oregon Department of Fish and Wildlife) established
Oregon 2020 surveys begin

2014
Willamette Bird Symposium, an annual day-long meeting celebrating birds, with short scientific presentations, begins

Principal Oregon Bird Publications and Editors

Alan L. Contreras and Martha Schmitt

Oregon Naturalist (1894–1898)
G. B. Cheney, 1894
A. B. Averill, 1895
John Martin, 1896–1898

Oregon Birds (and predecessor *SWOC Talk*)
The bibliographic history of *Oregon Birds* is somewhat complicated; users of the journal and its index should be aware of this. The change of name from *SWOC Talk* to *Oregon Birds* occurred in 1977 in the middle of volume 3. The last issues of *SWOC Talk* are vol. 3, no. 1, (Jan./Feb. 1977) to vol. 3, no. 3 (May/June 1977) pages continuously. With the change of name, volume numbering was discontinued, and all issues of *Oregon Birds* through volume 5 are paged individually. Aug./Sept. 1977, Oct./Nov. 1977, and Dec. 1977/Jan. 1978 were published as issue no. 1 to issue no. 3. With issue no. 4, Feb./Mar. 1978, volume numbering was resumed, continuous with *SWOC Talk* and issues 1–3, 1977 regarded as vol. 3, no. 4–6. The editors and indexer agreed that these issues would be so cited in the Index:

> Issue no. 1, Aug./Sept. 1977 as vol. 3, no. 4
> Issue no. 2, Oct./Nov. 1977 as vol. 3, no. 5
> Issue no. 3, Dec. 1977/Jan. 1978 as vol. 3, no. 6

Issue numbering was discontinued after issue no. 5, vol. 4, no. 2, April/May 1978. Only three issues, vol. 4, no. 1, Feb./Mar. 1978–; vol. 4, no. 3, June/July

1978, were published in 1978; vol. 4, no. 4, Aug./Sept./ 1978, vol. 4, no. 5, Oct. 1978 and vol. 4, no. 6, Dec. 1978 were published in early 1979. After vol. 5, no. 1, Feb. 1979, monthly designations were discontinued and the remaining issues of 1979 bear only volume, number and year citations.

[*The commentary above appeared originally with Martha Schmitt's index to* Oregon Birds, *volumes 3–5. It is reproduced here because it remains useful to anyone using* Oregon Birds.]

Oregon Birds, vol. 35 (2009) was issued as a single extra–large issue in April 2010. *Oregon Birds*, vol. 36 (2010) to date was issued as two full–color issues, except that OB 41(2) was not issued.

Oregon Field Ornithologists changed its name to Oregon Birding Association in 2012 after an advisory vote of the membership.

Oregon Birds began in 1975 as *SWOC TALK*, the newsletter of the Southern Willamette Ornithological Club (SWOC), based in Lane County. This group still meets but does not produce a publication. SWOC and its newsletter were founded by George A. "Chip" Jobanek of Walterville, Lane County. He was the first editor of what became Oregon's statewide bird publication.

The following list provides concordance between ST/OB volume numbers and years of issue. In some cases the actual year of issue was later than the titular year, especially with the early late-fall issues, which sometimes appeared after the first of the following year. *SWOC Talk* began life as a bimonthly newsletter.

The following list includes all editors of *Oregon Birds* and *SWOC TALK*.

Volume	Year	Editor
1	1975	George A. "Chip" Jobanek
2	1976	George A. "Chip" Jobanek through 2 (4); E. G. White-Swift and Alan Contreras, 2 (5) and 2 (6).
3	1977	E. G. White-Swift
4	1978	E. G. White-Swift through 4 (2); Alan Contreras 4 (3) through 4 (6).
5	1979	Alan Contreras 5 (1); Jim Carlson, Alan Contreras, and Steve Gordon 5 (2) through 5 (6).
6	1980	Steve Gordon (*Oregon Birds* became quarterly)
7	1981	Steve Gordon
8–10	1982–1984	Jim Carlson
11	1985	Dave Irons 11 (1); Owen Schmidt 11 (2) through 11 (4)

12–24	1986–1998	Owen Schmidt
25	1999	Ray Korpi, Alan Contreras, Mike Patterson, Reid Freeman
26	2000	Matt Hunter, Ray Korpi
27–30	2001–2004	Steve Dowlan
31	2005	Steve Dowlan, Jeff Harding
32–34	2006–2008	Jeff Harding
35–40	2009–2014	Alan Contreras (*Oregon Birds* became biannual, with field notes in spring and articles in fall.)
41–44	2015–2018	Hendrik Herlyn, Oscar Harper
45	2019	Selena Deckelmann, Linda Tucker Burfitt
46	2020	Diana Byrne, Linda Tucker Burfitt
47	2021	Linda Tucker Burfitt, Mike Williams

Journal of Oregon Ornithology (1993–1996)
Range Bayer, editor. Inactive but not closed.

Christmas Bird Count Editors for Oregon
1972–1974   Fred Zeillemaker
1974–1983   Larry McQueen
1983–1984   Tom Mickel and Larry McQueen
1984–1986   David Fix
1987–1988   Alan Contreras
1988–1991   Jeff Gilligan
1991–1996   Alan Contreras
1996–2016   Mike Patterson
2016–   Joel Geier

Presidents, Oregon Statewide Bird Clubs

Vjera E. Thompson

Northwest Ornithological Association
1894 Arthur Pope (founding president,
Northwestern Ornithological Association (NOA)
1895 William L. Finley
1896 William L. Finley
1897 William L. Finley

Oregon Birding Association and predecessor Oregon Field Ornithologists
1980 M. S. "Elzy" Eltzroth (founding president, Oregon Field
Ornithologists (OFO), February 2, 1980)
1981 M.S. "Elzy" Eltzroth
1982 M.S. "Elzy" Eltzroth
1983 Richard Palmer
1984 Richard Palmer
1985 Matt Hunter
1986 Matt Hunter
1987 Alan Contreras
1988 Alan Contreras
1989 Larry Thornburgh
1990 Bill Stotz
1991 David A. Anderson
1992 David A. Anderson
1993 Tim Shelmerdine

1994	Tim Shelmerdine
1995	George A. Jobanek
1996	Mike Patterson
1997	Mike Patterson
1998	Ray Korpi
1999	Ray Korpi
2000	Ray Korpi
2001	Paul T. Sullivan
2002	Mary Anne Sohlstrom (first woman president)
2003	Mary Anne Sohlstrom
2004	Mary Anne Sohlstrom
2005	Tim Shelmerdine
2006	Tim Shelmerdine
2007	Dave Tracy
2008	Dave Tracy
2009	Dan Gleason
2010	Jeff Harding
2011	Russ Namitz
2012	Russ Namitz
2013	Pamela Johnston
2014	Pamela Johnston
2015	Pamela Johnston
2016	Harv Schubothe
2017	Cathy Nowak
2018	Jimmy Billstine
2019	Diana Byrne
2020	Nagi Aboulenein
2021	Sarah Swanson
2022	Brodie Cass Talbott

Secretaries, Oregon Bird Records Committee

Alan L. Contreras

Alan L. Contreras (1978–1979)
Clarice Watson (1980–March 1989)
Tom Staudt (April 1989–April 1990)
Harry B. Nehls (May 1990–March 2015)
Treesa Hertzel (April 2015–present)

Repositories of Field Journals, Specimens, and Recordings from Oregon

Alan L. Contreras

Note: See VertNet for current information on specimens and sound recordings

CAS = California Academy of Sciences
MCZ = Museum of Comparative Zoology, Harvard University
MVZ = Museum of Vertebrate Zoology, University of California, Berkeley
OHS = Oregon Historical Society
OSU = Oregon State University
SDMNH = San Diego Museum of Natural History
SOU = Southern Oregon University
UO = University of Oregon Museum of Natural and Cultural History
UPS = University of Puget Sound
USNM = US National Museum (Smithsonian)
WFVZ = Western Foundation for Vertebrate Zoology

Location of journals/field notebooks of Oregon observers
Note: these institutions may hold only part of an observer's field notes. In some cases what is held relates only to collection of specimens; in other cases more general notes are available.

A. W. Anthony	WFVZ (1894–1897)
H. E. Anthony	USNM
Vernon Bailey	University of Wyoming, USNM

Charles Bendire	USNM
Herman Bohlman	OSU (photos), OHS
John Bovard	UO
Sidney Carter	SOU (nest and egg photos)
Edmund Currier	WFVZ
Overton Dowell Jr.	WFVZ
Merlin Eltzroth	OSU
William L. Finley	OHS, OSU
Grace McCormac French	OSU (Hist. of Science Archive)
Ira Gabrielson	WFVZ
W. E. Griffee	WFVZ
Henry Henshaw	OHS (final report)
Joseph Hicks	SOU
Stanley Jewett	USNM, some at UPS
Ruth Hopson Keen	UO
Urban Kubat	WFVZ
David B. Marshall	Retained by Marshall family
Cathy Merrifield	OSU (Hatfield MSC)
Olaus Murie	USNM
John E. Patterson	SOU
Morton E. Peck	USNM
Edward Preble	USNM
Hubert Prescott	UO (early years)
Albert G. Prill	OHS (papers)
Carl Richardson	SOU
H. H. Sheldon	USNM
Alfred C. Shelton	UO
Otis Swisher	SOU
Alex Walker	WFVZ

Notes

Introduction

1   Those interested in the intersection of Native life and birds are referred to the following sources. Amadeo Rea, *Wings in the Desert: A Folk Ornithology of the Northern Pimans* (Tucson: University of Arizona Press, 2008); Eugene Hunn, *Nch'i-WÄ¡na, The Big River: Mid-Columbia Indians and Their Land* (Seattle: University of Washington Press, 1990); Sonia Tidemann and Andrew Gosler, *Ethno-Ornithology: Birds, Indigenous Peoples, Culture and Society* (London: Earthscan, 2011), and Kailyn Chandler, Robert James, Samrat Clements, Moore; María Nieves Zedoño; Wendi Field Murray, *The Winged: An Upper Missouri River Ethno-ornithology* (Tucson: University of Arizona Press, 2017).

2   J. Michael Scott, Thomas W. Haislip Jr., and Margaret Thompson, "A Bibliography of Oregon Ornithology (1935–1970) with a Cross-referenced List of the Birds of Oregon," *Northwest Science* 46 (1972): 122–139. Mark Egger, *Bibliography of Oregon Ornithology: An Updating for the Years 1971-1977, with a Revised, Cross-Referenced List of the Birds of Oregon*, Oregon Field Ornith. Spec. Pub. No. 1 (1980).

3   As this book was being submitted, the servers holding the BirdNotes database were damaged. It is not clear whether the database will recover.

4   William Behle, *Utah Birds: Historical Perspectives and Bibliography*, Utah Museum of Natural History Occasional Publication No. 9 (1990).

Chapter 1: Origins and Types of Bird Study

1   For a history of this early period from a European perspective, see Erwin Stresemann, *Ornithology from Aristotle to the Present* (English trans., Cambridge, MA: Harvard University Press, 1975) and Paul Farber, *The Emergence of Ornithology as a Scientific Discipline: 1760-1850* (Dordrecht: Reidel, 1982).

2   M. Ralph Browning, *Rogue Birder* (Eugene: Oregon Review Books, 2018), 6.

3   Rosenberg is quoted in a story on bird declines that appeared in the Eugene, Oregon, *Register-Guard* online on Saturday, December 28, 2019.

4   Robert Pennock, *An Instinct for Truth: Curiosity and the Moral Character of Science* (Cambridge, MA: MIT Press, 2019), 42.

5   Marianne G. Ainley, "The Contribution of the Amateur to North American Ornithology," *Living Bird* 18 (1979–80): 167.

6   Farber, *Emergence of Ornithology*, 104.

7   Farber, 122.

8.   Stresemann, *Ornithology from Aristotle to the Present*, 351.

9.   Steve Howell, Ian Lewington, and Will Russell, *Rare Birds of North America* (Princeton, NJ: Princeton University Press, 2014), 2.

Chapter 2. Euro-American Land-based Expeditions, 1804–1859

1   See, for example, Theed Pearse, *Birds of the Early Explorers of the Northern Pacific* (Comox, BC: Theed Pearse, "The Close," 1968), an excellent self-published hardcover book that is mostly concerned with British Columbia and Alaska.

2   The quotation from Lewis and Clark's journal and the comment on it are from Ira Gabrielson and Stanley Jewett, *Birds of Oregon* (Corvallis: Oregon State College [now Oregon State University Press], 1940), 46–47.

3   See also Jesse D'Elia and Susan M. Haig, *California Condors in the Pacific Northwest* (Corvallis: Oregon State University Press, 2013).

4   Vic Coggins, in *Birds of Oregon: A General Reference*, ed. Marshall et al. (Corvallis: Oregon State University Press, 2003), 618–619.

5   John K. Townsend, *Narrative of a Journey Across the Rocky Mountains to the Columbia River*, annotated by George A. Jobanek (1839; Corvallis: Oregon State University Press, 1999). Much of the material on Townsend presented here comes from Jobanek's invaluable annotated edition of Townsend's memoir, including the extensive and detailed Introduction.

6   Lewis A. McArthur and Lewis L. McArthur, *Oregon Geographic Names*, 7th ed. (Portland: Oregon Historical Society, 2003).

7   M. Ralph Browning, "Type Specimens of Birds Collected in Oregon," *Northwest Science* 53, no. 2 (1979): 132.

8   Baptiste Dorion's role is noted in *Astoria: Astor and Jefferson's Lost Pacific Empire* (New York: Ecco, 2015), Peter Stark's history of the Astor expeditions.

9   George A. Jobanek, *An Annotated Bibliography of Oregon Bird Literature Published Before 1935* (Corvallis: Oregon State University Press, 1997).

10   Spencer F. Baird, J. Cassin, and G. Lawrence, *General Report upon the Zoology of the several Pacific Railroad Routes. Part II. Birds* (Washington, DC: Smithsonian Institution, 1858).

Chapter 3. The Contributions of Charles Emil Bendire

1   See Alan L. Contreras and Ulrich Hardt, eds., *Collected Poems of Ada Hastings Hedges* (Corvallis: Oregon State University Press, 2020).

2   Ira Gabrielson and Stanley Jewett, *Birds of Oregon* (Corvallis: Oregon State College [now Oregon State University Press], 1940), 53.

3   Thomas Brewer [Charles Bendire], "Notes on Seventy-nine Species of Birds Observed in the Neighborhood of Camp Harney, Oregon, Compiled from the Correspondence of Capt. Charles Bendire, 1st Cavalry," *Proceedings of the Boston Society of Natural History* 18 (1875): 153–168. Brewer compiled Bendire's correspondence for this article, but only added an introductory paragraph of his own, so the article

is often listed under Brewer's name, but internally in the original publication it is treated as having Bendire as the author.

4   Adapted and updated from M. Ralph Browning, Bendire's records of Red-shouldered Hawk (*Buteo lineatus*) and Yellow-bellied Sapsucker (*Sphyrapicus varius nuchalis*) in Oregon. *The Murrelet* 54, no. 3 (1973): 34–35.

5   Charles Bendire, "Notes on Some of the Birds Found in Southeastern Oregon, Particularly in the Vicinity of Camp Harney, from November, 1874, to January, 1877," *Proceedings of the Boston Society for Natural History* 19 (1877): 109–149.

6   George Willett, "Bird Notes from Southeastern Oregon and Northeastern California," *Condor* 21 (1919):196–206.

7   There is no other written record of what Chief Joseph said, and C. E. S. Wood is generally considered to have written the speech based on several things that Joseph said at the time. It is therefore not a translation in the usual sense of the term, but a creative interpretation.

8   Standard biographical treatments Jobanek used are J. C. Merrill, "In memoriam: Charles Emil Bendire," *Auk* 15 (1898): 1–6; C. Hart Merriam, "Charles Bendire," *Science* 5 (1897): 261–262; F. H. Knowlton, "Major Charles E. Bendire," *Osprey* 1 (1897): 87–90; and "Major Charles Bendire," *Forest and Stream* 48 (February 13, 1897). Bendire's acquaintance with C. E. S. Wood is further explored in Edwin R. Bingham, "Experiment in Launching a Biography: Three Vignettes of Charles Erskine Scott Wood," *Huntington Library Quarterly* 35 (1972): 221–239.

Chapter 4. Early Statehood, 1860–1900: Travelers and Local Lists

1   Robert Shufeldt, "Contributions to Avian Paleontology: Studies of the Fossil Birds of the Oregon Desert." *Auk* 13 (1913): 36. See also his "On a Collection of Fossil Birds from the Equus Beds of Oregon," *American Naturalist* 25 (1891): 359–362; "Tertiary Fossils of North American Birds," *Auk* 8 (1891): 365–368; "A Study of the Fossil Avifauna of the Equus Beds of the Oregon Desert," *Journal of the Academy of Natural Sciences of Philadelphia* 9 (1892): 389–425; and "Review of the Fossil Fauna of the Desert Region of Oregon, with a Description of Additional Material Collected There," *Bulletin of the American Museum of Natural History* 32 (1913): 123–178.

2   Loye Miller, *Lifelong Boyhood: Recollections of a Naturalist Afield* (Berkeley: University of California Press, 1950).

3   Loye Miller, *Journal of First Trip of University of California to John Day Fossil Beds of Eastern Oregon*, ed. J. Arnold Shotwell, Bulletin No. 19, University of Oregon Museum of Natural History, [1899] 1972.

4   Loye Miller, "The Birds of the John Day Region, Oregon," *Condor* 6 (1904): 100–106. A unique aspect of this publication is that an effort was made to count birds in the same areas in 1983. See Brian Sharp, "Avifaunal Changes in Central Oregon since 1899," *Western Birds* 16, no. 2 (1985): 63–70.

5   See Barbara Stein, *On Her Own Terms: Annie Montague Alexander and the Rise of Science in the American West* (Berkeley: University of California Press, 2001).

6   Edgar A. Mearns, "A Partial List of the Birds of Fort Klamath, Oregon, collected by Lieutenant Willis Wittich, USA, with annotations and additions by the collector," *Bulletin of the Nuttall Ornithological Club* 4 (1879): 161–166, 194–199.

7   J. C. Merrill, "Notes on the Birds of Fort Klamath, Oregon," *Auk* 5 (1888): 139–146, 251–262, 357–366.

8   See the ornithological report in the *Annual Report of the US Geographical Survey West of the 100th Meridian for 1879*. See also George A. Jobanek, "Early Records of the White-faced Ibis in Oregon," *Oregon Birds* 13 (1987): 210–215.

9   The quote from Henshaw's memoir appears in Ira Gabrielson and Stanley Jewett, *Birds of Oregon* (Corvallis: Oregon State College, 1940), 53–54.

10   A. W. Anthony, "Field Notes on the Birds of Washington County, Oregon." *Auk* 3 (1886): 161–172.

11   George A. Jobanek, "Dubious Records in the Early Oregon Bird Literature," *Oregon Birds* 20, no. 1 (1994): 3–23.

12   There was persistent discussion of the presence of this Common Crane in the birding community at the time, but it was said to be in a closed area of Sauvie Island and no photos were ever seen.

Chapter 5. The Northwestern Ornithological Association

1   Most of this narrative is constructed from readings of the *Oregon Naturalist*, 1894 to 1898. Lacking access to this little journal (most of it is now available online), readers can discover this literature in Jobanek's *Annotated Bibliography of Oregon Bird Literature Published Before 1935* (Corvallis: Oregon State University Press, 1997). Jobanek has also written of Pope's Oregon bird list in "Towards a Revised Bibliography of Oregon Ornithology prior to 1935," *Oregon Birds* 13, no. 1 (1987): 56–59, and discussed the dubious species on the list in "Dubious Records in the Early Oregon Bird Literature," *Oregon Birds* 20, no. 1 (1994): 3–23. The chapter on the Northwestern Ornithological Association (NOA) is adapted and revised from Jobanek's article that originally appeared in *Oregon Birds* 24, no. 1 (1998): 12.

2   Bernard Bretherton's coastal notes appear in Range Bayer's "1884–1923 Oregon Coast Bird Notes in Biological Survey Files," *Studies in Oregon Ornithology*, no. 1 (Newport, OR: Gahmken Press, 1986), and his "Records of Bird Skins Collected Along the Oregon Coast," *Studies in Oregon Ornithology*, no. 7 (Newport, OR: Gahmken Press, 1989).

3   Some sources say that Hoskins attended Harvard, but the family was Quaker, and Haverford, a Quaker college listed in other sources, seems more likely to be correct. The title County Judge, as applied to Hoskins, meant that he was the presiding officer of what would now be called the county board of commissioners.

Chapter 6. The Legacy of William L. Finley and Herman T. Bohlman

1   For a book-length treatment of Finley's life based on photographs and interviews with Finley's family and friends, see Worth Mathewson, *William L. Finley, Pioneer Wildlife Photographer* (Corvallis: Oregon State University Press, 1986). Two addi-

tional biographies are currently underway, one by Joe Blakely and one by a member of Finley's extended family.

2   See Chapter 5, "The Northwestern Ornithological Association," by George A. Jobanek.

3   Lord thanks A. W. Anthony, Herman Bohlman, and Ross Nicholas, all NOA members who had help form the John Burroughs Club, in the foreword of his book on Oregon birds. William Rogers Lord, *A First Book upon the Birds of Oregon and Washington* (Boston: Heintzemann Press, 1902).

4   Sources indicate different dates for the beginning of the John Burroughs Club/Oregon Audubon Society. Audubon Society of Portland gives the date for the John Burroughs Club founding as 1898. Frances S. Twining in 1927 gave the date as 1900 after consulting the John Burroughs Club minutes book. However, the minutes book seems to have subsequently been lost. Lord dates his arrival in Oregon as spring of 1899 in his book on Oregon and Washington birds. Herman Bohlman's treasurer's book for the Oregon Audubon Society shows a transfer of funds from JBC to Oregon Audubon Society as of 1901. See Tom McAllister, "Our First 50 Years—1902–1952," Audubon Society of Portland; Frances S. Twining, "Early History of Bird Societies," typescript noted *Sunday Oregonian*, October 9, 1927, Box 27, MSS 2990, Oregon Audubon Society Records, Oregon Historical Society Research Library, Portland, Oregon (hereafter cited as OAS); Account Book 1901–1930, OAS. Mabel Osgood Wright to Gertrude Metcalf, Feb. 2, 1902, Correspondence, General 1902–1957, OAS. OAS was separate from the National Audubon Society until it agreed to become a chapter under certain conditions in 1966.

5   Preston E. Anderson, "Ornithological Photography: An Appreciation of Two Earnest Workers," *Camera Craft* 14 (1907): 122.

6   William L. Finley, "Catching Birds with a Camera," *The Condor* 3, no. 6 (1901): 137–139. See also William L. Finley, "Hunting Birds with a Camera: A Record of Twenty Years of Adventure in Obtaining Photographs of Feathered Wild Life in America," *National Geographic*, August 1923, 160.

7   Finley, "Catching Birds with a Camera." See also Finley, "Hunting Birds with a Camera," 160.

8   Finley included details about how he and Bohlman worked together in his many published pieces. See also Anderson, "Ornithological Photography: An Appreciation of Two Earnest Workers"; Preston E. Anderson, "Bird Studies and Pictures from Life with a Camera: William Lovell Finley, Herman T. Bohlman," *The Craftsman* 8 (1905): 613–622; Mathewson, *William L. Finley, Pioneer Wildlife Photographer*.

9   William L. Finley, *American Birds* (New York: Scribners, 1907), 3.

10 Finley, *American Birds*, 57–63.

11 Mathewson, 33–34. The name Ross Nicholas is sometimes given as Ron as Mathewson did, or Nichols. In the Herman Bohlman Lecture Notes collection at the Oregon Historical Society (OHS), Bohlman writes that it was Ross Nicholas along with Bohlman, Ellis Hadley, and Finley on the 1901 trip to Netarts. Bohlman's writing is not totally clear on the name, but it looks more like ss than n. Ross Nicholas also gets thanked along with Bohlman by Rev. Lord in Lord's 1901 book and by A. R. Wood-

cock in his 1902 compilation, for the use of his and Bohlman's notes. He also is in the Audubon records in Bohlman's treasurer's book as a member of John Burroughs/ Audubon and did a fundraising lecture for the organization with Bohlman. Oregon State University (OSU) and OHS still have "two unidentified men" listed with the photos they have of the group setting out from Dayton and in Netarts.

12 William L. Finley, "Among the Sea Birds of the Oregon Coast," *The Condor* 4, no. 3 (1902): 53–57.

13 William L. Finley, "Among the Sea Birds off the Oregon Coast, Part I." *The Condor* 7, no. 5 (1905): 119–127; William L. Finley, "Among the Sea Birds off the Oregon Coast, Part II." *The Condor* 7, no. 6 (1905): 161–169.

14 "State Reports: Oregon," *Bird-Lore* 7, no. 1 (January–February 1905): 105.

15 Archival materials indicate Finley was still working to pass matriculation exams at UC Berkeley in late 1899 and winter 1900. William L. Finley Papers, 1899–1946, MSS Finley, Series 7: Personal Materials, 1899–circa 1922, Oregon State University Libraries. Also available at digitalcollections.ohs.org.

16 "Twenty-Second Congress of the American Ornithologists' Union," *The Auk* 22 (1905): 74–75; "New Bird Photographer," *Evening Star*, Washington, DC, November 22, 1905.

17 Finley is first listed as a paying member in July 1902. Account Book 1901–1930, OAS.

18 The fundraiser brought in $23.75 for the organization. Account Book 1901–1930, OAS.

19 Finley to Grinnell, May 16, 1905, and July 17, 1905, Folder-Finley, Box 7, Joseph Grinnell Papers, BANC MSS C-B 995, The Bancroft Library, University of California, Berkeley (hereafter Joseph Grinnell papers, BL); William L. Finley, "The Cruise of Two Camera Hunters," *Pacific Monthly* 23 (1910): 632–641; "State Reports: Oregon," *Bird-Lore* (1905): 336–342.

20 Finley, "The Cruise of Two Camera Hunters"; William L. Finley, "Among the Gulls on Klamath Lake," *The Condor* 9, no. 1 (1907): 12–16; William L. Finley, "Among the Pelicans," *The Condor* 9, no. 2 (1907): 35–41; William L. Finley, "The Grebes of Southern Oregon," *The Condor* 9, no. 4 (1907): 97–101.

21 The merger between the Portland-based group and the Astoria group was not finalized at the corporate level until 1909, according to the treasurer's notes kept for Portland by Herman Bohlman.

22 The Oregon Audubon chapter provided $400 toward the cost of the trip. Entries for May 2, 1908, and April 10, 1909, Account Book 1901–1930, OAS.

23 Report of William L. Finley, *Bird-Lore* 10 (1908): 291–295.

24 Ibid.; William L. Finley, "The Trail of the Plume Hunter," *Atlantic Monthly* 106 (1910): 373–379.

25 Fish and Game Commission Board Minutes, 1910–1915, Department of Fish and Wildlife, State of Oregon Archives.

26 Editorial, *Oregon Sportsman* 2, no. 3 (March 1914): 1–2.

27 A number of authors have written about the competing demands on the Klamath Basin and Malheur Lake, and Finley's advocacy for the refuges. See J. Doug Foster,

"Necessary Co-existence: Lower Klamath and Tule Lake Wildlife Refuges, Part I and Part II," in Shaw Historical Library, *Wings That Fill the Sky: America's First Waterfowl Refuge* (Klamath Falls, OR: Shaw Historical Library, 2008); Stephen R. Mark, "Conflict and Compromise: William L. Finley and the Revival of Lower Klamath Lake," in *Wings That Fill the Sky*; Robert M. Wilson, *Seeking Refuge: Birds and Landscapes of the Pacific Flyway* (Seattle: University of Washington Press, 2010); Nancy Langston, *Where Land & Water Meet: A Western Landscape Transformed*, Weyerhaeuser Environmental Books (Seattle: University of Washington Press, 2003); Stephen Most, *River of Renewal: Myth & History in the Klamath Basin* (Portland: Oregon Historical Society Press, 2008).

28   Langston, *Where Land & Water Meet*, 62, 86–87; Wilson, *Seeking Refuge*, 62; "Finley Ousted as State Biologist," *Morning Oregonian*, December 18, 1919.
29   William L. Finley and Arthur Newton Pack, "Passing of the Marshlands," digitized from film reels, William L. Finley Papers, 1899-1946 (MSS Finley), Oregon State University Special Collections and Archives Research Center, Corvallis, Oregon. https://oregondigital.org
30   Langston, *Where Land & Water Meet*, 88.
31   Ibid.; Wilson, *Seeking Refuge*, 70–71.
32   Langston, 89.
33   "Malheur Refuge an Aid to Both Birds and Farmers," typescript, William L. Finley Papers, 1899–1946, MSS Finley, Series 1: Manuscripts, circa 1910–1942, Oregon State University Library.
34   For Finley's campaigning for cleaning the Willamette, see James V. Hillegas-Elting, *Speaking for the River: Confronting Pollution on the Willamette, 1920s–1970s* (Corvallis: Oregon State University Press, 2018).

Chapter 7. Introduced Birds in Oregon

1   This chapter is adapted and abridged from George A. Jobanek, "Bringing the Old World to the New: The Introduction of Foreign Songbirds into Oregon," *Oregon Birds* 13, no. 1 (1987): 59. Citations to sources are available in the original article.
2   The other large acclimatization society of the nineteenth century, the American Acclimatization Society, was organized in the late 1870s in order to introduce into New York state all of the species of birds mentioned by Shakespeare. In 1890 and 1891, the society succeeded in establishing the European Starling as a breeding bird in New York City. The starling spread steadily across the continent, increasing its range and numbers every year. By December 1943, it reached Oregon, a lone bird at Malheur National Wildlife Refuge headquarters. The officers of the Portland Song Bird Club probably dreamed of achieving just such a spectacular success with their introductions.
3   See George A. Jobanek, "History of the Bobolink in Oregon," *Oregon Birds* 20, no. 2 (1994): 50.
4   Oregon Audubon Society (now Portland Audubon Society), entries for February 24 and May 4, 2012, in record book No. 1, which covers September 11, 1909–October 20, 1923.

5   Joseph E. Evanich Jr., "Introduced Birds of Oregon," *Oregon Birds* 12, no. 3 (1986): 156.

6   Wire's history of the Game Commission is available online from the Oregon Department of Fish and Wildlife.

Chapter 8. Government Surveys and Academic Study, 1901–1939

1   Jenks Cameron, *The Bureau of Biological Survey, its History, Activities and Organization* (Baltimore: Johns Hopkins Press, 1919; New York: Arno Press reprint, 1974), 2.

2   Ira Gabrielson and Stanley Jewett, *Birds of Oregon* (Corvallis: Oregon State College [now Oregon State University Press], 1940). Additional resources exist for the Oregon Coast. See Range Bayer's "1884–1923 Oregon Coast Bird Notes in Biological Survey Files," *Studies in Oregon Ornithology*, No. 1 (Newport, OR: Gahmken Press, 1986), and his "Records of Bird Skins Collected Along the Oregon Coast," *Studies in Oregon Ornithology*, No. 7 (Newport, OR: Gahmken Press, 1989). These compilations are extremely thorough and detailed. They include annotations by Bayer explaining obscure details, for example the fact that early maps used by coastal observers mistakenly called Yaquina Head by the name Cape Foulweather, which is a different feature many miles to the north. Records included in these resources include those by lighthouse keeper James Langlois at Cape Blanco, for whom the town of Langlois in Curry County is named; Bernard Bretherton, the Briton who spent much of his career on the Oregon Coast including the Coquille Lighthouse; extensive material from Overton Dowell at Mercer Lake, Lane County; and Frank Plummer at Yaquina Head. There is a small amount of material from interior Oregon included.

3   Range Bayer, "Transcription of Vernon Bailey's field notes for his 1909 trip to Lincoln Co., Coos Co., and Curry Co., Oregon with notes about Biological Survey records," *Journal of Oregon Ornithology* 5 (1996): 614–625.

4   Ernest Thompson Seton with E. A. Preble, *The Arctic Prairies. A Canoe Journey of 2,000 Miles in Search of the Caribou; Being an Account of a Voyage to the Region North of Aylmer Lake* (New York: Scribner's, 1911).

5   Noah K. Strycker, ed., *Early Twentieth Century Ornithology in Malheur County, Oregon.* Oregon Field Ornithologists Special Publication No. 18 (2003).

6   Morton Peck also published a little-known short story and some poetry in *The Book of the Bardons* (Boston: Gorham Press, 1925).

7   Alan Contreras and Kenneth C. Parkes, "First Confirmed Record of Veery for Malheur County, Oregon," *Oregon Birds* 21, no. 3 (1995): 77.

8   Adapted in part from George A. Jobanek, "Some Comments on Alfred Cooper Shelton," *Oregon Birds* (SWOC Talk) 1, no. 3 (1977): 22. See also Jobanek's biographical summary in Noah Strycker's annotated reprint of *Shelton, Oregon, Field Ornithologists Special Publication* Number 14 (2002).

9   Alfred Cooper Shelton, *A Distributional List of the Land Birds of West Central Oregon* 14, no. 4 (1917) in the New Series of University of Oregon bulletins. An annotated reprint was issued by Oregon Field Ornithologists; see Noah K. Strycker, ed., *OFO Special Publication No. 14* (2002).

10 George A. Jobanek, "Searching for the Tree Vole: An Episode in the 1914 Biological Survey of Oregon," *Oregon Historical Quarterly* 89, no. 4 (1988): 369.

11 Remsen and Herman's comments are extracted from e-mails to Contreras in November 2019.

12 The material on Sutton is adapted from "Sutton in Oregon" by George A. Jobanek, *Oregon Birds* 13, no. 1 (1987): 52.

13 Letter from Johnson A. Neff to Otis Swisher, April 8, 1971. Swisher collection, Hannon Library, Southern Oregon University.

14 From the history of the Oregon Department of Fish and Wildlife, online on March 1, 2021. This history timeline is detailed and worth reading.

15 Frannk B. Wire's history can be found ont he Oregon Department of Fish and Wildlife page, A Brief History of the Oregon State Game Commission, 1938.

Chapter 9. Oregon Bird Books, 1901–1939

1 Some people do not realize that the 1901 edition of Lord's *Birds of Oregon* exists and refer to the 1902 edition as the "first edition." This is not correct. The problem is exacerbated by some rather confusing labeling by the publishers of the 1902 and 1913 editions and by Lord's introductory note in the 1913 edition, which says, "When this book was first issued in 1902. . . ." This refers to the first *revised* edition published by Heintzemann, not the first edition to bear the title. The first edition is a much smaller book with a meadowlark, not a grosbeak, on the cover.

2 A. R. Woodcock, *Annotated Checklist of the Birds of Oregon*, No. 68 in Oregon Agricultural College's publication series for its Agricultural Experiment Station, 1902.

3 In 1993, Contreras asked the Oregon Geographic Names Board to name Mt. Bailey in southern Oregon after Florence and Vernon to honor their biological work in Oregon. The board agreed. This was possible because the original name "Mt. Bailey" resulted from an erroneous transcription of the name Mt. Baldy; therefore, it was not named after anyone named Bailey. Now it is.

4 Ralph Hoffmann, *Birds of the Pacific States* (Boston: Houghton Mifflin, 1927), 123.

Chapter 10. Ira Gabrielson, Stanley Jewett, and . . . *Birds of Oregon*

1 Ira Gabrielson and Stanley Jewett, *Birds of Oregon* (Corvallis: Oregon State College, 1940). Reprinted by Dover (1970) under the deceptive title *Birds of the Pacific Northwest*.

2 There is more material available on Gabrielson than Jewett because Gabrielson wrote his memoirs through the year 1966 based on a daily diary and field notes. The memoirs have not been published. Marshall had one of several copies in existence, the location of which is now unknown. He made extensive use of the memoirs here as well as miscellaneous obituaries and other materials. Other information came from Gabrielson memoria materials assembled by Henry M. Reeves during a visit to the Patuxent Wildlife Research Center library and Wildlife Management Institute in 1985, and his own remembrances. Durward L. Allen, "Ira N. Gabrielson, 1889–

1977," *Wildlife Society Bulletin* 6, no. 2 (1978): 113–115; Henry M. Reeves and David B. Marshall, "In Memoriam: Ira Noel Gabrielson," *Auk* 102 (1985): 865–868.

3   Stanley Jewett and Ira Gabrielson, *Birds of the Portland Area, Oregon* (Berkeley, CA: Cooper Ornithological Club, 1929).

4   T. H. McAllister Jr. and D. B. Marshall, 1945. "Summer Birds of Fremont National Forest," *Auk* 62 (1945):176–189.

5   Stanley G. Jewett, Walter P. Taylor, William T. Shaw, and John W. Aldrich, *Birds of Washington State* (Seattle: University of Washington Press, 1953).

6   Range Bayer saw and copied some of Jewett's loose pieces of paper used in the preparation of *Birds of Oregon* during a visit to the University of Puget Sound Museum of Natural History in the 1980s.

Chapter 11. The Emergence of the Active Amateur, 1901–1960

1   "120 Years of American Education: A Statistical Portrait" (US Department of Education, 1993): 7.

2   Donald A. McCrimmon and Alexander Sprunt IV, *The Amateur and North American Ornithology, Proceedings of a Conference* (Ithaca, NY: National Audubon Society and Cornell University, 1978).

3   For more detail about the early life of Alex Walker, see Charles P. Crutchett, "Pioneer Prairie Ornithologists: Alex Walker," *South Dakota Bird Notes* (publication of South Dakota Ornithologist's Union) 6, no. 4 (December 1954): cover, 57–58, Range D. Bayer, "Incomplete Listing of Oregon Bird Records in A. C. Bent's Life History Series," *Journal of Oregon Ornithology* 5 (1996): 626–659; G.D.A. [Gordon Dee Alcorn], "In Memoriam: Alexander Walker 1890–1975," *Murrelet* 56, no. 3 (1975): 24–25; Range D. Bayer and Reed W. Ferris, "Reed Ferris' 1930–1943 Bird Banding Records and Bird Observations for Tillamook County, Oregon," *Studies in Oregon Ornithology*, no. 3 (1987); Harold C. Smith, "Early Biological Surveys of Oregon," *Oregon State Game Commission Bulletin* 16 (August 1961): 3–5, 7.

4   Alex Walker, "Some Birds of Central Oregon," *Condor* 19, no. 4 (1917): 131–140.

5   Alex Walker, "Notes on Some Birds from Tillamook County, Oregon," *Condor* 26, no. 5 (1924): 180–182.

6   Kenneth M. Walker, "Bullock's Oriole on the Oregon Coast," *Murrelet* 21, no. 2 (1940): 47.

7   Anonymous, obituary in the *Statesman Journal* (Salem, Oregon), October 18, 2016.

8   Alex Walker, "The Starling Reaches the Pacific," *Condor* 51, no. 6 (1949): 271.

9   For more about Walker's days at the Museum and later life, see Anonymous, "Alex Walker 1974," in *Tillamook History: Sequel to Tillamook Memories*, ed. Lila V. Cooper Boge and R. Evart (Tillamook, OR: Tillamook Pioneer Association, 1975), 7; Anonymous, "Alex Walker, Pioneer Museum Curator, Dies," *Headlight-Herald* (Tillamook, OR), August 20, 1975, p. 2.

10   For more about Ferris's time in Tillamook County, see Range D. Bayer, "Records of Bird Skins Collected along the Oregon Coast," *Studies in Oregon Ornithology* No. 7, 1989; Range D. Bayer and Reed W. Ferris, "Reed Ferris' 1930–1943 Bird Banding

Records and Bird Observations for Tillamook County, Oregon," *Studies in Oregon Ornithology* No. 3 (Newport, OR: Gahmken Press,1987); Range D. Bayer and Reed W. Ferris, "Return Rates of Terrestrial Birds Banded at Beaver, Tillamook County, Oregon," *Oregon Birds* 14, no. 1 (1988): 51–54; Range D. Bayer and Reed W. Ferris, "The Yearly Cycle of Common Murres along the Oregon Coast," *Oregon Birds* 14, no. 2 (1988): 150–152.

11 Reed W. Ferris, "Eight Years of Banding of Western Gulls," *Condor* 42, no. 4 (1940): 188–197, and Reed W. Ferris, G. D. Sprot, and M. C. Sargent, "Pacific Gull Color Banding Project," *Condor* 41, no. 1 (1939): 38.

12 Reed W. Ferris, "A Grinnell Water-Thrush in Oregon," *Condor* 35, no. 2 (1933): 80.

13 Bayer and Ferris, "Reed Ferris' 1930–1943 Bird Banding Records and Bird Observations for Tillamook County, Oregon."

14 Worth Mathewson, *William L. Finley, Pioneer Wildlife Photographer* (Corvallis: Oregon State University Press, 1986).

15 Gerry Hysmith, personal communication to Range Bayer from Tillamook Pioneer Museum, 1986.

16 French's notebooks have never been fully accessioned by OSU and need further study. A detailed summary of their content is available as "A Guide to the Bird Notes of Grace McCormac French of Yamhill County, Oregon" (Range Bayer, *Studies in Oregon Ornithology* No. 2 [Newport: Gahmken Press, 1986], 40 pp.). See also an excerpt with migration dates: Range Bayer, "Grace French's Arrivals and Departures of Birds in Yamhill County, Oregon," *Oregon Birds* 14, no. 3 (1988): 257–259.

17 For further information about Dowell's early years before he became a bird collector, see Anonymous, "Siuslaw High School Gets Unique Bird Collection," *Siuslaw News* (Florence, OR), January 31, 1963, p. 1; Anonymous, "Death Claims Two Local Pioneers," August 8, p. 1, *Siuslaw News*, August 8, 1963, p. 1; Anonymous, "Dowell Sells Large Ranch," *Siuslaw Oar* (Florence, OR), November 29, 1946, p. 1; Anonymous, "Death Claims Local Pioneer," *Siuslaw News*, November 1, 1962, p. 1; Stephanie Finucane and Jeannine Rowley, "Heceta House: A Historical and Architectural Survey," *Studies in Cultural Resource Management*, vol. 3, Waldport Ranger District, Siuslaw National Forest, US Forest Service-USDA, Pacific Northwest Region, 1980; Anonymous, "Oregon Fish and Game Commissioners, Regular Deputy Game Wardens, Regular Fish Wardens," *Oregon Sportsman* 3, no. 3 (1915): 214.

18 Stanley G. Jewett, "Directions for Preparing Scientific Specimens of Large and Small Mammals, Birds, Birds' Stomachs for Economic Investigations, Birds' Nests and Eggs, Fish and Reptiles," *Oregon Fish and Game Commission Bulletin* no. 1 (Salem, OR: State Printing Dept., 1914).

19 For further information about Dowell's collecting years, see Western Foundation of Vertebrate Zoology (WFVZ), Unpublished Field Notes of Overton Dowell Jr., Camarillo, California, 1987; Harold C. Smith, "Early Biological Surveys of Oregon," *Oregon State Game Commission Bulletin* 16 (August 1961): 3–5, 7; Ira N. Gabrielson and Stanley G. Jewett, *Birds of Oregon*, Studies in Zoology no. 2 (Corvallis: Oregon State University Monographs, 1940); Alfred C. Shelton, *A Distributional List of the*

Land Birds of West Central Oregon, University of Oregon Bulletin 14, no. 4 (1917);
Range D. Bayer, "1884–1923 Oregon Coast Bird Notes in Biological Survey Files,"
Studies in Oregon Ornithology No. 1 (1986); Range D. Bayer, "Records of Bird Skins
Collected Along the Oregon Coast," *Studies in Oregon Ornithology* No. 7 (1989);
Stanley G. Jewett, "Some New Bird Records from Oregon," *Condor* 44 (1942):
36–37; J. Michael Scott and Harry B. Nehls, "First Oregon Records for Thick-billed
Murre," *Western Birds* 5, no. 4 (1974): 137.

20 Stanley G. Jewett,. "Upland Plover (Bartramia longicauda) in Oregon," *Auk* 47, no. 1
(1930): 78.

21 Stanley G. Jewett, "Notes on the Dowell Bird Collection," *Condor* 32 (1930): 123–
124.

22 J. Michael Scott, Letter of July 8, 1970, to Richard Whitmore, Vice Principal of
Siuslaw High School, acknowledging that Overton Dowell's "Ornithological Col-
lection" (including his field notes) that had been at Siuslaw High is in the Natural
History Museum of the OSU Dept. of Zoology.

23 Dowell's Cassin's Kingbird is discussed in S. Jewett, "Some New Bird Records from
Oregon," *Condor* 44 (1942): 36–37, and in Ralph Browning, "'Notes on the Hypo-
thetical List of Oregon Birds," *Northwest Science* 48 (1974): 166. Thanks to Range
Bayer for providing a copy of Dowell's original notes, which are held at the Western
Foundation for Vertebrate Zoology.

24 Tom McCamant, "Early Rogue Valley Records," *Oregon Birds* 3, no. 5 (1977): 17.
See also Bing Francis, "Tom McCamant: 67 years of Bird Watching," *Oregon Birds* 7,
no. 1 (1981): 45.

25 *Sunday Oregonian*, October 10, 1920: 41.

26 Ira N. Gabrielson, Portland, Oregon, report in The Season section in *Bird-Lore* 24
(1922): 104.

27 Tom McAllister, *Our First Fifty Years, 1902–1952* (typescript provided by Portland
Audubon Society, undated, but after 2001).

28 See George A. Jobanek, "The Reliability of Dr. Albert G. Prill," *Oregon Birds* 19, no.
2 (1993): 44, for a more detailed account of Prill's life and activities.

29 Those interested in Reiher's life are referred to these principal sources. Linda Flaxel,
1978, "A Tribute to Hilda [Reiher]," *Cape Arago Tattler* (Cape Arago Audubon Society), 2, no. 1 (1978): 3–5; Anonymous, "[Obituary] Hilda Reiher," *The World*
(Coos Bay, OR), November 8, 1976, p. 2; Anonymous, "[Remembrance] Hilda
Reiher," *Audubon Warbler* (publication of Audubon Society of Portland), 40 (December 1976): 5; Hilda Reiher [note quoted in Field Notes], *Audubon Warbler* 38
(March 1974): 6; Hilda Reiher, "Notes from North Bend," *Audubon Warbler* 12, no.
8 (1949): 2; Hilda Reiher, "Christmas Bird Count: North Bend," *Audubon Warbler*
13. no. 5 (1950): 8–9; Hilda Reiher, "Christmas Bird Count: North Bend," *Audubon
Warbler* 14, no. 6 (1951): 9–10; Anonymous, "Warbles," *Audubon Warbler* 3, no. 3
(1939): 2. In addition, Reiher is mentioned in the *Coos Bay World* newspaper a number of times.

30 The course was probably the correspondence class developed by John Bovard and

later taught by Ralph Huestis using a somewhat smaller custom-made text.

31 Olive Barber, *Birds of Coos County* (self-published, 1938), 8 pp. (available from Oregon State Library, Salem). Olive Barber is remembered today primarily in Olive Barber Drive, a road along the hills of Eastside, a part of the City of Coos Bay, and as author of *Lady and the Lumberjack*, 1953, a memoir of her life married to a logger. See also Anonymous, "Olive Barber," *The State Reference: Capitol's Who's Who for Oregon* (Portland, OR: Capitol Publishing Company, 1948–1949), 42.

32 Alan Contreras, Birds of Coos County, Oregon, Status and Distribution, OFO Special Publication No. 12 (Cape Arago Audubon Society in Cooperation with Oregon Field Ornithologists, 1998).

33 Hilda Reiher, unpublished materials donated by Reiher's family after her death to the library of Southwestern Oregon Community College (SWOCC). Range Bayer photocopied "Bird Records of First Sightings of Unusual Species or Numbers" recorded from 1932 to 1974-Hilda Reiher-North Bend, OR. This was a typed, single sheet. There were also bird notes in a black spiral notebook that he only partially compiled into a three typed page compilation for arrival dates. He sent copies of the photocopied typed page and three-page compilation to Alan Contreras in 1997, and some notes were included in Contreras's *Birds of Coos County*.

34 Alan Contreras, "New Historic Records of Anna's Hummingbird from Oregon," *Western Birds* 30 (1999): 214.

35 A good example of Keen's botanical knowledge can be found in *Crater Lake Nature Notes* 34-3, vol. 13 (October 1947), available online at the Crater Lake website.

36 The Keen bird notes from the 1940s were found by Nancy Brown of Portland, who provided them to Contreras for this project. Paul Adamus undertook the evaluation of records and placed those deemed appropriate into eBird. Bill Sullivan and George Jobanek assisted with locating some sites in Lane County mentioned in Keen's notes.

Chapter 12. Teaching about Birds

1 Herb Wisner died at age ninety-nine on February 20, 2022, as this book was in production.

2 Gregory A. Green, in *Birds of Oregon: A General Reference*, ed. Marshall et al. (Corvallis: Oregon State University Press, 2003), 317–318.

3 Personal communication from Luke Suchoski, 2021.

4 Personal communication from David Bailey, 2020.

5 Gordon Gullion, "Birds of the Southern Willamette Valley, Oregon," *Condor* 53, no. 3 (May–June 1951): 129–149.

6 Stanley Gordon Jewett and Ira Noel Gabrielson in *Birds of the Portland Area, Oregon*, Pacific Coast Avifauna No. 19 (Berkeley, CA: Cooper Ornithological Club, 1929), mention finding twelve cuckoos on June 8, 1923, along the Columbia, by far the largest number reported anywhere in the state before or since. Apparently cuckoos were common in 1924 and 1925 also. We know of no comparable period in Oregon history.

Chapter 13. Development and Limits of the Citizen Scientist

1   In 2021, the American Ornithological Society changed the name of *Auk* to *Ornithology* and the name of *Condor* to *Ornithological Applications.*

2   M. S. Eltzroth, President's Message, *Oregon Birds* 6, no. 1 (1980): 2.

3   A. Contreras, "Wampole's 1957–1959 Annotated Checklist of Birds of Coos Bay, Oregon," *Journal of Oregon Ornithology* 5 (1996): 545–557.

4   Thomas J. Crabtree and Harry B. Nehls, "A Checklist of the Birds of Oregon," *Western Birds* 12 (1981): 145–156.

5   Steve Summers and Craig Miller, *Preliminary Draft: Oregon County Checklists and Maps,* Oregon Field Ornithologists (OFO) Special Publication No. 7, February 1993.

Chapter 14. Oregon's Expanding Avifauna

1   Phillips is quoting Davies, *Ibis* 119 (1977): 557 (1977).

2   See also George A. Jobanek, "The Growth of the Oregon Birdlist, 1906 through 1995," *Oregon Birds* 23, no. 4 (1997): 120.

3   Email from Mike Denny to Contreras, November 2019.

4   See *Oregon Birds* 19, no. 2 (1993): 40–44 for several related articles regarding the Patterson Yellow Rail reports.

5   W. E. Griffee, "First Oregon Nest of Yellow Rail (*Coturnicops noveboracensis*)," *Murrelet* 25, no. 2 (1944): 29.

Chapter 15. Expanding Knowledge of Oregon's Seabirds

1   Matthew Hunter and Greg Gillson, "The Future of Oregon's Oceanic Birding," *Oregon Birds* 23, no. 2 (1997): 40–46.

2   Ira Gabrielson and Stanley Jewett, *Birds of Oregon* (Corvallis: Oregon State University, 1940).

3   Gerald Sanger, "The Seasonal Distribution of Some Seabirds off Washington and Oregon, with Notes on their Ecology and Behavior" (Seattle: University of Washington, Department of Oceanography, 1965).

4   Michael Scott, personal communication with Nolan M. Clements, 2019.

5   Tom Crabtree, personal communication with Nolan M. Clements, 2020.

6   Paul Sullivan, "Pelagic Fall Migration in Oregon Waters," *Oregon Birds* 14, no. 2 (1988): 134–144.

7   Tim Shelmerdine, personal communication with Nolan M. Clements, 2020.

Chapter 16. Atlases, Surveys, and Counts

1   G. G. Beck, A. R. Couturier, C. M. Francis, and S. Leckie, *North American Ornithological Atlas Committee Handbook: A Guide for Managers on the Planning and Implementation of a Breeding Bird Atlas Project* (Port Rowan, Ontario, Canada: Bird Studies Canada, 2018).

2   P. R. Adamus, K. Larsen, G. Gillson, and C. Miller, *Oregon Breeding Bird Atlas* (Eugene: Oregon Field Ornithologists, 2001, CD-ROM).

3   A list of publications using CBC data is maintained by the National Audubon Society on its CBC website. The web address as of this publication is https://www.audubon.org/christmas-bird-count-bibliography.

4   Jim Carlson, Lister's Corner, *Oregon Birds* 7, no. 1 (1981): 53.

Chapter 17. The Long Wingspan of David B. Marshall

1   Some sources indicate that Dave Marshall was also a Fellow of the American Ornithologists' Union, which is a separate category. The American Ornithological Society, successor to the AOU, has clarified that this is not correct. The national membership list for the major bird organizations uses an "F" to indicate members of Association of Field Ornithologists, and Dave's entry says "F 1989." It appears that biographical notes about Dave picked this up as an indicator that he became a Fellow in 1989, an error that Contreras repeated in his memorial for Marshall (Alan L. Contreras, "In Memoriam: David B. Marshall, 1926–2011," *The Auk* 129, no. 1 [January 1, 2012]: 177–178). We take this opportunity to correct the record.

2   This segment is condensed from a series of articles that Dave Marshall wrote for *Oregon Birds*.

3   Tom McAllister (1926–2018). See the chapters by McAllister in *Edge of Awe: Experiences of the Malheur-Steens Country* (Corvallis: Oregon State University Press, 2019) for his view of growing up as a young birder in Portland.

Chapter 18. *The Life History of Birds of Oregon: A General Reference*

1   As a footnote to history, Dave was traveling with Judith Ramaley and Barbara Holland from Portland State University, both of whom remain friends, which led to my visiting Vermont and Australia to enjoy the birds found there.—Alan Contreras

2   For more background on the *Birds of Oregon* project, see David B. Irons, "Harry Nehls Receives OFO's First 'Lifetime Service Award,'" *Oregon Birds* 33, no. 1 (2007): 1–3. (In this article, Dave Irons includes Dave Marshall's tribute to Harry Nehls, which includes his contribution to BOGR.) David Marshall, preface and acknowledgments, in *Birds of Oregon: A General Reference*, by D. Marshall, M. Hunter, and A. Contreras (Corvallis: Oregon State University Press, 2003). David Marshall, *Memoirs of a Wildlife Biologist* (self-published, 2008; available through Portland Audubon Society). David Marshall, "Seventy Years of a Young Birder's Experience in Oregon," *Oregon Birds* 36, no. 1 (2010): 2–4, extracted herein.

Chapter 19. The Internet Age

1   *Oregon Birds* 21, no. 3 (1995).

2   Greg Gillson personal communication with Vjera Thompson, 2019.

3   *Oregon Birds* 20, no. 2 (1994): 61.

4   *Oregon Birds* 21, no. 2 (1995): 42.

5   Rich Hoyer personal communication with Vjera Thompson, 2019.

6   Dennis Arendt personal communication with Vjera Thompson, 2019.

7   Greg Gillson personal communication with Vjera Thompson, 2019.

8 Bill Tice, in *Oregon Birds* 22, no. 3 (1996): 76.

9 Bill Tice, in *Oregon Birds* 22, no. 3 (1996):76.

10 Joel Geier personal communication with Vjera Thompson, 2019.

11 Chart of eBird growth: https://ebird.org/news/ebird-2018-year-in-review

12 http://nwbackyardbirder.blogspot.com/2012/03/birdseye-bird-log-killer-app-for.html

13 Personal communication from Van Remsen to Contreras, November 21, 2019.

14 W. D. Robinson, C. Partipilo, T. A. Hallman, K. Fairchild, and J. P. Fairchild, "Idiosyncratic Changes in Spring Arrival Dates of Pacific Northwest Migratory Birds," PeerJ 7 (2019): e7999, https://doi.org/10.7717/peerj.7999.

15 OBOL email dated 04/01/2017.

16 OBOL email dated 01/09/2002.

17 OBOL email dated 01/11/2002.

18 OBOL email dated 12/18/2002.

19 Announcement of profile pages: https://ebird.org/news/profilepages/.

20 https://www.ecaudubon.org/birding-locations.

21 *Oregon Birds* 36, no. 1 (2010): 15.

22 http://www.iusedtohatebirds.com/p/the-5mr-2019-challenge.html.

23 OBOL email dated 10/14/2019.

Chapter 20. Hopes, Predictions, and Expectations

1 Steven G. Herman, 2002. "Wildlife Biology and Natural History: Time for a Reunion," *Journal of Wildlife Management* 66, no. 4 (2002): 933–946.

Contributors

PAUL ADAMUS of Corvallis, Oregon, managed the Oregon Breeding Bird Atlas project. He also oversees Oregon's Breeding Bird Survey routes.

RANGE BAYER of Newport, Oregon, is retired and was editor of the *Journal of Oregon Ornithology* and of the series Studies in Oregon Ornithology.

DANNY BYSTRAK works for the Bird Banding Lab in Maryland.

M. RALPH BROWNING of Medford, Oregon, is retired from the Division of Birds, US Museum (Smithsonian).

NOLAN M. CLEMENTS of La Grande, Oregon, recently graduated from Oregon State University.

ALAN L. CONTRERAS (standing) of Eugene, Oregon, is retired from a career in higher education oversight and has written and edited several books on Northwest ornithology. (**STEVE HERMAN**, seated.)

JEFF FLEISCHER worked for fifteen years as an assistant refuge manager on six National Wildlife Refuges in the West and finished his career working for the US Postal Service.

HENDRIK HERLYN of Corvallis, Oregon, is a biologist, translator, and senior author of the *Handbook of Oregon Birds* (Oregon State University Press, 2013).

STEVE HERMAN was a professor at The Evergreen State College in Washington State.

MATT HUNTER (right) of Roseburg, Oregon, is an independent wildlife ecologist, co-editor of *Birds of Oregon* (Oregon State University Press, 2003), and taught at Umpqua Community College. (**DAVE MARSHALL**, left.)

GEORGE A. JOBANEK of Eugene, Oregon, is author of *An Annotated Bibliography of Oregon Ornithology Published Before 1935* (Oregon State University Press, 1997) and editor of the Oregon State University Press reprint of John K. Townsend's *Narrative of a Journey Across the Rocky Mountains to the Columbia River*.

The late **TOM MCALLISTER** worked for many years as outdoor writer for the *Oregon Journal* and *Oregonian* newspapers.

The late **DAVID B. MARSHALL** served for many years with the US Fish and Wildlife Service and was senior editor of *Birds of Oregon* (Oregon State University Press, 2003).

CAREY MYLES of Portland, Oregon, wrote her master's thesis (Portland State University) on the career of William L. Finley.

M. CATHY NOWAK of Union, Oregon, is a biologist working for the Oregon Department of Fish and Wildlife in La Grande.

MIKE PATTERSON of Astoria, Oregon, is a biologist and bird-bander.

NOAH K. STRYCKER from Creswell, Oregon, is a writer and naturalist working mainly in the Arctic and Antarctic regions.

VJERA E. THOMPSON of Eugene, Oregon, is a reviewer for the eBird database and was an early volunteer with Birdnotes.

TERESA WICKS of Burns, Oregon, is a biologist employed by the Portland Audubon Society who works primarily in Harney County.

~

JOSEPH E. EVANICH JR. was a Portland-based artist and a student at Eastern Oregon University. He illustrated *Birder's Guide to Oregon, Birds of Northeast Oregon*, and other publications and was a principal illustrator for *Oregon Birds* magazine. He died from complications of AIDS at age thirty-three in 1993.

BARBARA B. C. GLEASON is a Eugene-based artist whose work has appeared in *Birds of Lane County, Oregon, Birds From the Inside Out*, and other publications.

SUSAN LINDSTEDT illustrated *Birds of Malheur National Wildlife Refuge* and other publications. A former resident of La Grande, she now lives in southwestern Idaho.

Index